California Science

Science Content Support
Grade 4

Harcourt
SCHOOL PUBLISHERS

Visit *The Learning Site!*
www.harcourtschool.com

Copyright © by Harcourt, Inc.

All rights reserved. No part of this publication may be reproduced or transmitted in any form or by any means, electronic or mechanical, including photocopy, recording, or any information storage and retrieval system, without permission in writing from the publisher.

Permission is hereby granted to individuals using the corresponding student's textbook or kit as the major vehicle for regular classroom instruction to photocopy Copying Masters from this publication in classroom quantities for instructional use and not for resale. Requests for information on other matters regarding duplication of this work should be addressed to School Permissions and Copyrights, Harcourt, Inc., 6277 Sea Harbor Drive, Orlando, Florida 32887-6777. Fax: 407-345-2418.

HARCOURT and the Harcourt Logo are trademarks of Harcourt, Inc., registered in the United States of America and/or other jurisdictions.

Printed in the United States of America

ISBN-13: 978-0-15-352283-3
ISBN-10: 0-15-352283-6

2 3 4 5 6 7 8 9 10 022 15 14 13 12 11 10 09 08 07 06

If you have received these materials as examination copies free of charge, Harcourt School Publishers retains title to the materials and they may not be resold. Resale of examination copies is strictly prohibited and is illegal.

Possession of this publication in print format does not entitle users to convert this publication, or any portion of it, into electronic format.

Contents

Getting to Know Your Textbook . . . . CS vii

Getting Ready for Science

Lesson 1 What Are Investigation Tools?
Vocabulary Power . . . . CS 1
Study Skills . . . . CS 2
Quick Study . . . . CS 3

Lesson 2 What Are Investigation Skills?
Vocabulary Power . . . . CS 5
Study Skills . . . . CS 6
Quick Study . . . . CS 7

Lesson 3 How Do Scientists Use Graphs?
Vocabulary Power . . . . CS 9
Study Skills . . . . CS 10
Quick Study . . . . CS 11

Lesson 4 What Is the Scientific Method?
Vocabulary Power . . . . CS 13
Study Skills . . . . CS 14
Quick Study . . . . CS 15

Unit 1 • Electricity and Magnetism

Lesson 1 What Is Static Electricity?
Vocabulary Power . . . . CS 17
Study Skills . . . . CS 18
Quick Study . . . . CS 19

Lesson 2 What Makes a Circuit?
Vocabulary Power . . . . CS 21
Study Skills . . . . CS 22
Quick Study . . . . CS 23
Extra Support, Designing and Building Circuits . . . . CS 25

Lesson 3 What Are Magnetic Poles?
Vocabulary Power . . . . CS 29
Study Skills . . . . CS 30
Quick Study . . . . CS 31

Lesson 4 How Can You Detect a Magnetic Field?
Vocabulary Power . . . . CS 33
Study Skills . . . . CS 34
Quick Study . . . . CS 35

Lesson 5 What Makes an Electromagnet?
　　Vocabulary Power . . . . .CS 37
　　Study Skills . . . . .CS 38
　　Quick Study . . . . .CS 39

Lesson 6 How Are Electromagnets Used?
　　Vocabulary Power . . . . .CS 41
　　Study Skills . . . . .CS 42
　　Quick Study . . . . .CS 43

Lesson 7 How Is Electrical Energy Used?
　　Vocabulary Power . . . . .CS 45
　　Study Skills . . . . .CS 46
　　Quick Study . . . . .CS 47

Unit 2 • Energy for Life and Growth

Lesson 1 What Are Producers and Consumers?
　　Vocabulary Power . . . . .CS 49
　　Study Skills . . . . .CS 50
　　Quick Study . . . . .CS 51

Lesson 2 What Are Decomposers?
　　Vocabulary Power . . . . .CS 53
　　Study Skills . . . . .CS 54
　　Quick Study . . . . .CS 55

Lesson 3 What Are Food Chains and Food Webs?
　　Vocabulary Power . . . . .CS 57
　　Study Skills . . . . .CS 58
　　Quick Study . . . . .CS 59

Lesson 4 How Do Living Things Compete for Resources?
　　Vocabulary Power . . . . .CS 61
　　Study Skills . . . . .CS 62
　　Quick Study . . . . .CS 63

Unit 3 • Ecosystems

Lesson 1 What Makes Up an Ecosystem?
Vocabulary Power .... CS 65
Study Skills .... CS 66
Quick Study .... CS 67

Lesson 2 What Affects Survival?
Vocabulary Power .... CS 69
Study Skills .... CS 70
Quick Study .... CS 71

Lesson 3 What Is Interdependence?
Vocabulary Power .... CS 73
Study Skills .... CS 74
Quick Study .... CS 75

Lesson 4 What Are Microorganisms?
Vocabulary Power .... CS 77
Study Skills .... CS 78
Quick Study .... CS 79

Unit 4 • Rocks and Minerals

Lesson 1 How Are Minerals Identified?
Vocabulary Power .... CS 81
Study Skills .... CS 82
Quick Study .... CS 83
Extra Support, Identifying Minerals .... CS 85

Lesson 2 How Are Rocks Identified?
Vocabulary Power .... CS 89
Study Skills .... CS 90
Quick Study .... CS 91

Lesson 3 What Is the Rock Cycle?
Vocabulary Power .... CS 93
Study Skills .... CS 94
Quick Study .... CS 95

Unit 5 • Waves, Wind, Water, and Ice

Lesson 1 What Causes Changes to Earth's Surface?
    Vocabulary Power . . . . CS 97
    Study Skills . . . . CS 98
    Quick Study . . . . CS 99

Lesson 2 What Causes Weathering?
    Vocabulary Power . . . . CS 101
    Study Skills . . . . CS 102
    Quick Study . . . . CS 103

Lesson 3 How Does Moving Water Shape the Land?
    Vocabulary Power . . . . CS 105
    Study Skills . . . . CS 106
    Quick Study . . . . CS 107

Vocabulary

    Vocabulary Games . . . . CS 109
    Vocabulary Cards . . . . CS 113

Name _____

Date _____

Getting to Know Your Textbook

Welcome to *Science* by Harcourt School Publishers. You can look forward to an exciting year of discovery. Your textbook has many features that can help you learn science this year. Use this scavenger hunt to learn more about it.

1. What animal is on the cover of your book? _____

   Name one fact about the animal. _____

2. Find the unit called "Getting Ready for Science." Name one of the science tools you will use this year. _____

3. Name two of the steps of the Scientific Method, found in the same unit.

   _____

4. What are the three handbooks in the back of your book?

   _____

5. The Big Idea is what you will understand by the end of the unit. What's the "Big Idea" for Unit 5? _____

6. In order to understand the Big Idea, you will be answering Essential Questions. The Essential Questions are also the titles of the lessons. What is one of the Essential Questions for Unit 3?

   _____

7. What Reading Focus Skill is used in Lesson 2 of Unit 2?

   _____

8. What is the first word in the glossary? _____

9. What is the last word listed in the index? _____

10. Find an activity that you would like to try on one of the Make Connections pages. Write its title and page number. _____

Science Content Support **CS vii**

Name _____

11. List the numbers of the California Standards that you will learn about in Unit 2, Lesson 1. _____

12. Write the title and page number of an Investigate you would like to try. What's the Investigation Skill Tip? _____

13. Write the title and page number of an Insta-Lab you would like to try.

_____

14. Find a California on Location feature that you find interesting. What's the location? _____

15. Find the postcard at the beginning of Unit 3. Who's it from? Where was it sent from? _____

16. You can use the Vocabulary Preview to learn new science words from each lesson. The preview shows you how to say, or pronounce, the term. Find the Vocabulary Preview for Unit 2, Lesson 1. List two vocabulary words that you will learn. What photos are shown to help you understand those words? _____

17. The words in the Vocabulary Preview also appear in the text of your book. They are highlighted in yellow and are used in a way that helps to explain their meanings. Find the two words from the item above in the text and list the page number each appears on. _____

18. Find a California Fast Fact that you find interesting. Write its title and page number. _____

19. Write the name of a person featured in one of the People features.

_____

20. Write three new things you expect to learn about this year.

_____

_____

_____

CS viii   Science Content Support

Name _____

Date _____

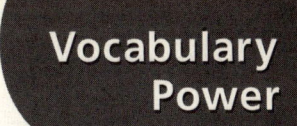

Introductory Unit, Lesson 1

Lesson 1—What Are Investigation Tools?

Context Clues

Read each sentence. Think about the meaning of the underlined word. Write the meaning on the line.

1. Instead of using body parts to measure distance, today we use a standard measure, such as a meter.

   _____

   _____

2. Joel examined the tiny insect up close using his microscope.

   _____

   _____

3. Brianna used a spring scale to measure the weight of the turtle she found.

   _____

   _____

4. Juanita carefully squeezed the forceps to lift the prickly burdock.

   _____

   _____

5. John used a tape measure instead of a ruler to measure the circumference of his table.

   _____

   _____

6. Geologists use hand lenses to look at rock crystals.

   _____

   _____

Use with Introductory Unit.   Science Content Support

Name _____

Date _____

Study Skills

Introductory Unit, Lesson 1

Lesson 1—What Are Investigation Tools?

Use Visuals

Visuals can help you better understand and remember what you read.

- Photographs, illustrations, diagrams, charts, and maps are different kinds of visuals.
- Many visuals have titles, captions, or labels that help readers understand what is shown.
- Visuals often show information that appears in the text, but in a different way.

As you read this lesson, look closely at the visuals and the text that goes with them. Answer the questions in the checklist.

Checklist for Visuals

✔	What kind of visual is shown? _____
✔	What does the visual show? _____
✔	What does the visual tell you about the topic? _____
✔	How does the visual help you understand what you are reading? _____

Science Content Support

Use with Introductory Unit.

Name _____
Date _____

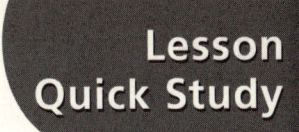

Introductory Unit, Lesson 1

Lesson 1—What Are Investigation Tools?

1. **Investigation Skill Practice–Measure**

    Jonathan needs to add 10 drops of soap to the bubble mixture he is making. The tools in his science lab include a measuring cup, beaker, graduate, and dropper. Which tool should he select to accurately measure this volume of liquid? Why?

    _____
    _____

2. **Reading Skill Practice–Main Idea and Details**

    Read the selection. Underline the main idea. List at least two details about the main idea.

    Christina wanted to measure the temperature of the chocolate that she was melting. She knew that she could use the numbers on a thermometer to measure how warm the liquid was. Christina carefully placed the thermometer in the melted chocolate. As the liquid inside the thermometer got warmer, it expanded and rose up in the tube. Then she read the number closest to the top of the liquid to discover the temperature of the melted chocolate.

    _____
    _____
    _____
    _____

Use with Introductory Unit.   (page 1 of 2)   Science Content Support   CS 3

Name _____

Science Concepts

3. In the chart below, write an object in the second column telling what you can measure or observe with the investigation tool in the first column.

Tool	What I Can Measure With It
hand lens	patterns on a beetle
dropper	
forceps	
magnifying box	
spring scale	
thermometer	
ruler	
flexible measuring tape	
microscope	
graduate	

4. How would measurements be different today if people still used body parts to measure distances?

_____
_____
_____
_____

Science Content Support   Use with Introductory Unit.

Name _____
Date _____

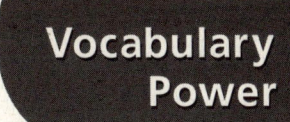

Introductory Unit, Lesson 2

Lesson 2—What Are Investigation Skills?

Word Family Activity

Complete the sentences with the correct form of the word.

1. observe/observations

   Noun: Jose recorded his _____ of leaf shapes in his sketchbook.

   Verb: Regan wanted to _____ the lunar eclipse from her backyard.

2. inference/infer

   Verb: Gregory could _____ that the larger rabbit ate more carrots than the smaller rabbit.

   Noun: Charlene based her _____ on what she observed as she watched the frogs in her pond.

3. prediction/predict

   Noun: My _____ is that the goose's eggs will hatch on Wednesday.

   Verb: The weatherperson could _____ snow tomorrow.

4. estimation/estimate

   Verb: Ron decided to _____ the height of the tallest giraffe in the zoo.

   Noun: Paula's _____ was that the giraffe was more than 17 feet tall.

5. hypothesis/hypothesize

   Noun: Marsha needed a microscope to test her _____.

   Verb: I will _____ that the holes should be at least 6 centimeters for the bean bag toss.

Use with Introductory Unit.   Science Content Support   CS 5

Name _____
Date _____

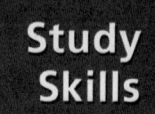

Study Skills

Introductory Unit, Lesson 2

Lesson 2—What Are Investigation Skills?

Make an Outline

An outline is a good way to record main ideas and details.

- Topics in an outline are shown by Roman numerals.
- Main ideas about each topic are shown by capital letters.
- Details about each main idea are identified by numbers.

As you read this lesson, write down the topics, main ideas, and details. Use the information to complete the outline below.

_____
_____
_____
_____

What Are Investigation Skills?

I. Observe and Infer
   A. Inquiry Skill
      1. can help you think like a scientist
   B. Observations
      1. information from the senses
      2. can record with notes and drawings
   C. Inference
      1. untested interpretation
      2. based on observations
      3. based on what you know

II. Predict
   A. Prediction
      1. _____
      2. _____
      3. _____
      4. _____

Science Content Support   Use with Introductory Unit.

Name _____

Date _____

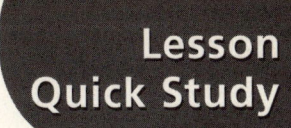

Introductory Unit, Lesson 2

Lesson 2—What Are Investigation Skills?

1. Investigation Skill Practice–Use Models

In the Investigate you made a model of a building. What would you plan so you could double the height of your building? How would your model compare to an actual building?

_____
_____
_____
_____
_____
_____

2. Reading Skill Practice–Main Idea and Details

Read the selection. Underline the main idea. List at least three details about the main idea.

Brian used inquiry skills to predict that the cocklebur was designed to travel far from the plant. He observed that the burr had hooked bristles. He already knew that the burr clung to his dog's fur when they hiked. In his experience, the plant was widely distributed. Based on this pattern of events, Brian took an educated guess about the design of the cocklebur.

_____
_____
_____
_____
_____

Use with Introductory Unit.   (page 1 of 2)   Science Content Support   CS 7

Name _____

Science Concepts

3. In the chart below, write how each investigation skill might be used in the second column.

Investigation Skill	How It Might Be Used
observe	the color of a hummingbird's feather
infer	
predict	
compare	
classify	
use numbers	
time/space relationships	
use models	
measure	
plan an experiment	
display data in a graph	
communicate	

4. Suppose you were helping your friends make homemade ice cream. What investigation skills might you use?

_____
_____
_____
_____

Name _____

Date _____

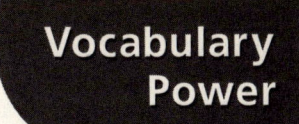

Introductory Unit, Lesson 3

Lesson 3—How Do Scientists Use Graphs?

Explore Word Meanings

Write the words from the box to complete each sentence. Use clues in the sentences and glossary definitions to help you select the correct words.

| interpretation | scale | axis |
| table | circle | |

1. I used a _____ on my bar graph to show the size of units marked in centimeters.

2. Her _____ of the data was that most students preferred ice cream for dessert.

3. You can transfer the data from a _____ to a bar graph to easily compare the information.

4. A _____ graph shows a whole that is made up of parts.

5. You can look down or left on a line graph to see each point's value on an _____ .

Use with Introductory Unit.   Science Content Support

Name _____

Date _____

Study Skills

Introductory Unit, Lesson 3

Lesson 3—How Do Scientists Use Graphs?

Understand Vocabulary

Using a dictionary can help you learn new words that you find as you read.

- A dictionary shows all the meanings of a word and tells where the word came from.
- You can use a chart to list and organize unfamiliar words that you look up in a dictionary.

As you read this lesson, look up familiar words in the dictionary. Add them to the chart below. Fill in each column to help you remember the word's meaning.

> **scale** (skāl) **n. 1.** A series of tones going up or down in pitch. **2.** A series of spaces marked by lines and used to measure distances or to register something. **3.** The size of a picture, plan, or model of a thing compared to the size of the thing itself. **4.** Size in comparison. **5.** A rule by which something can be measured or judged. (from Middle English *scale* "ladder")

Word	Syllables	Origin	Definition
scale	SKAYL	Middle English	A series of spaces marked by lines and used to measure distances or to register something

CS 10 Science Content Support   Use with Introductory Unit.

Name _____
Date _____

Lesson Quick Study

Introductory Unit, Lesson 3

Lesson 3—How Do Scientists Use Graphs?

1. Investigation Skill Practice – Graph

Look at the graph. What do the labels on the y-axis and the x-axis show? What does the line graph show?

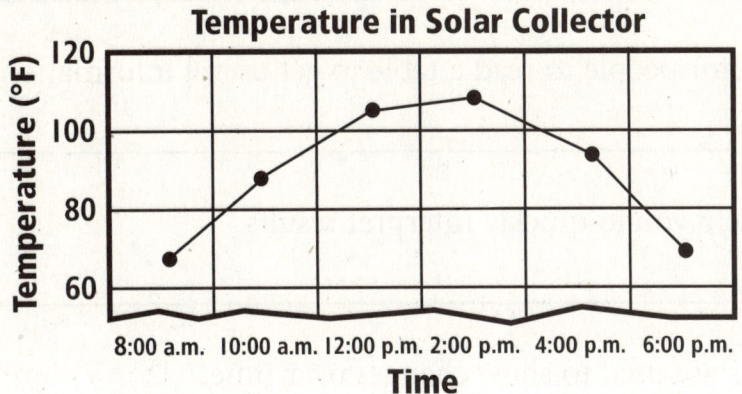

_____

_____

_____

_____

2. Reading Skill Practice – Main Idea and Details

Read the selection. Underline the main idea. Circle at least two details about the main idea.

    Ms. Shea's fourth-grade class needed to make a graph for a science project. They decided to display their data in a circle graph to easily compare whether students were right- or left-handed. First they took a poll. They counted 40 students in all and recorded their data in a table. Next, they displayed their data in a circle graph. The graph showed that 10 percent of the students were left-handed and 90 percent of the students were right-handed.

Name _____

Science Concepts

3. **Each statement below has a mistake. Rewrite the sentence correctly.**

   Scientists graph information to help them make observations.
   _____
   _____

   It is difficult for people to read a table to get useful information.
   _____
   _____

   Tables can help you to quickly interpret results.
   _____
   _____

   Circle graphs are used to show changes over time.
   _____
   _____

   There are interpretations along each axis.
   _____
   _____

   A line graph shows data as a whole made up of parts.
   _____
   _____

4. **Walter is studying the weight of his pet guinea pig over a 6-month period. His guinea pig is on a diet, and the veterinarian recommended that Walter weigh his pet once a month. What would be a good way for Walter to display his data for the vet? Why?**

   _____
   _____
   _____

Name _____

Date _____

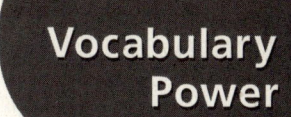

Introductory Unit, Lesson 4

Lesson 4—What Is the Scientific Method?

Context Clues

Read each sentence. Think about the meaning of the underlined word. Write the meaning on the line.

1. The <u>scientific method</u> can help you to test ideas.

   _____
   _____

2. Sandy needed to test her <u>hypothesis</u> that she could rollerskate faster on wood floors than on carpet because wood has lower rolling resistance.

   _____
   _____

3. Only one <u>variable</u> would be tested in Roger's experiment.

   _____
   _____

4. Sara got the same result for each <u>trial</u> in her experiment.

   _____
   _____

5. Jim <u>drew conclusions</u> after he completed his experiment.

   _____
   _____

6. After experimenting with the straws, the students wrote a <u>report</u>.

   _____
   _____
   _____

Use with Introductory Unit.   Science Content Support   CS 13

Name _____
Date _____

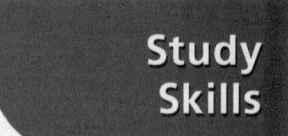

Introductory Unit, Lesson 4

Lesson 4—What Is the Scientific Method?

Preview and Question

Identifying main ideas in a lesson and asking questions about them can help you find important information.

- To preview a lesson, read the lesson title and the section titles. Look at the pictures, and read their captions. Try to get an idea of the main topic, and think of questions you have.
- Read to find the answers to your questions. Then recite, or say, the answers aloud. Finally, review what you have read.

As you read this lesson, fill in the chart and practice reading, reciting, and reviewing.

What Is the Scientific Method?

Preview	Questions	Read	Recite	Review
Using the five steps of the scientific method to test ideas	How can you test your hypothesis?	✓	✓	✓

CS 14 Science Content Support   Use with Introductory Unit.

Name _____
Date _____

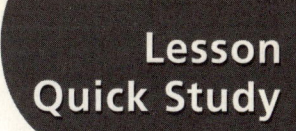

Introductory Unit, Lesson 4

Lesson 4—What Is the Scientific Method?

1. **Investigation Skill Practice – Experiment**

   Suppose you wanted to test whether animals with dark-colored fur warmed up more quickly than animals with light-colored fur on a sunny day. Describe an experiment to test your hypothesis.

   _____
   _____
   _____
   _____

2. **Reading Skill Practice – Main Idea and Details**

   Read the selection. Underline the main idea. List at least two details about the main idea.

   Yvonne shared the result of her experiment so others could double-check her work. By double-checking Yvonne's work, others should get similar results when they repeated her investigation. If they identified any mistakes, others could build new ideas on reliable knowledge. Yvonne's findings would allow others to learn from her experiment.

   _____
   _____
   _____
   _____

Use with Introductory Unit.   (page 1 of 2)   Science Content Support   CS 15

Name _____

Science Concepts

3. Check (✔) the statements below that agree with the information found in the lesson.

   _____ There are seven steps in a scientific method that help scientists test ideas.

   _____ You can begin to test an idea by forming a hypothesis.

   _____ You should only change one variable on the trials of your experiment.

   _____ If you get different results for each trial, your measurements were done well.

   _____ You should record everything you observe as you conduct your experiment.

   _____ The only way to record the results of your experiment is in numbers.

   _____ The first step of a scientific method is to conduct an experiment.

   _____ Another person should be able to get similar results when repeating your experiment.

   _____ You can share your findings in a written or oral report.

   _____ Charts, graphs, and diagrams help explain your conclusions.

4. **Suppose you need to know which type of candy will melt the fastest. You predict that the chocolate bar would melt fastest. Your test results support this. A friend predicts that the toffee would melt fastest. What step of the scientific method would help your friend to learn from your experiment? Why?**

   _____

   _____

   _____

   _____

Name _____

Date _____

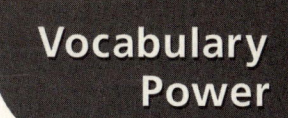

Unit 1, Lesson 1

Lesson 1—What Is Static Electricity?

A. Explore Word Meanings

Write the word or words from the box that matches the definitions below. Use all of the words once.

electric field	neutral	repel	static electricity

_____ 1. the buildup of electric charges in one place

_____ 2. the area around electric charges where electric forces can act

_____ 3. to push away from something

_____ 4. positive and negative charges cancel each other when matter has a balance of charges

B. Context Clues

Write the term from the box that completes each sentence.

attract	positively charged	electric charge

1. You can tell that that two balloons _____ each other when one balloon pulls toward the other.

2. When two balloons are _____ they will push away from one another.

3. An _____ may be positive or negative.

Use with Unit 1.

Name _____

Date _____

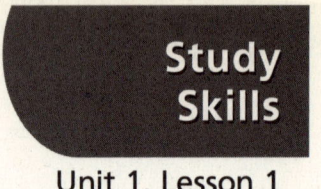

Unit 1, Lesson 1

Lesson 1—What Is Static Electricity?

Use a K-W-L Chart

A K-W-L chart can help you focus on what you already know about a topic and what you want to learn about it.

- Use the K column to list what you already know about electricity.
- Use the W column to list what you want to learn about electricity.
- Use the L column to list what you have learned about electricity from your reading.

As you read this lesson, complete the K-W-L chart below.

Electricity		
What I **K**now	What I **W**ant to Learn	What I **L**earned
• I know I get a shock after walking on carpet in my socks.	• What makes the shock? • How do clothes get static electricity in the dryer?	• _____
• _____ • _____ • _____	• _____ • _____ • _____	• _____ • _____ • _____

CS 18 Science Content Support

Use with Unit 1.

Name _____
Date _____

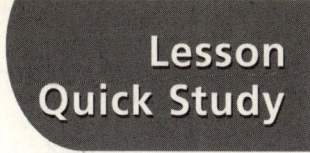

Unit 1, Lesson 1

Lesson 1—What Is Static Electricity?

1. Investigation Skill Practice – Infer

Nadine rides in a car with soft fabric seats. When Nadine stops the car, she slides on the seats to get out. She pushes the metal car door shut with her bare hand. She gets a shock! The shock is made by static electricity. What can you infer about the interaction between the car seat, the car door, and Nadine?

_____
_____
_____
_____

2. Reading Skill Practice – Cause and Effect

Read the selection. Describe a cause and effect related to static electricity.

Mario washed his clothes and put them in the dryer. When he took the clothes out of the dryer, some socks were stuck to his shirts. What happened? Simple! As clothes rotate in the dryer, they rub against each other. Negative charges move around from one sock or shirt to another. Some clothes get a positive charge. Other clothes get a negative charge. This build up of charges produces static electricity. Opposite charges attract each other. The clothes that built up negative charges stick to the clothes with positive charges.

_____
_____
_____
_____

Name _____

Science Concepts

3. In the chart below, tell which balloons would be attracted and repelled by each balloon listed.

	This balloon would attract _____	This balloon would repel _____
balloon A negative charge (−)		
balloon B positive charge (+)		
balloon C neutral charge		
balloon D negative charge (−)		
balloon E positive charge (+)		
balloon F neutral charge		

4. Name one way that a balloon could become negatively charged.

_____

_____

_____

CS 20   Science Content Support   Use with Unit 1.

Name _____
Date _____

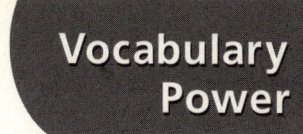

Unit 1, Lesson 2

Lesson 2—What Makes a Circuit?

A. Graphic Organizer – Explore Word Meanings

Fill in the graphic organizer with the missing information.

Type of Circuit	How It Works	Description
electric circuit		It may contain any number of batteries, wires, or bulbs.
	an electric circuit with only one path for current	When one bulb breaks, all bulbs go out.
parallel circuit		When one bulb breaks, others stay lit.
		Low resistance and high current produce a lot of heat. These can be dangerous.

B. Context Clues

Fill in the blanks in the paragraph with the correct word from the box.

electric current	resistance	short circuit

Julio found a lamp in the barn. He loved the red glass shade. He looked at the wires. They looked very old. Julio felt them get hot when he plugged the lamp in. He knew that the _____ was too low. That's why the wires got hot. He was determined to fix the wires so they would carry _____ to the bulb. His father unplugged the lamp, then took it apart. He fixed the wires so they didn't touch. He did not want to _____ the lamp when he plugged it back in. When the repairs were done, the lamp lit up with a beautiful red glow.

Use with Unit 1.　　　　　　　　　　　Science Content Support　　CS 21

Name _____
Date _____

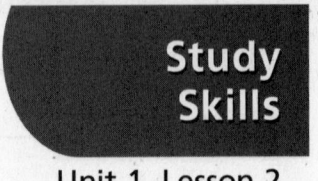

Unit 1, Lesson 2

Lesson 2—What Makes a Circuit?

Make an Outline

An outline is a good way to record main ideas and details.

- Topics in an outline are shown by Roman numerals.
- Main ideas about each topic are shown by capital letters.
- Details about each main idea are identified by numbers.

As you read this lesson, remember to pay attention to the topics, main ideas, and details. Use the information to add to the outline below. One section has been done for you.

What Makes a Circuit?

I. Moving Charges

   A. An object becomes charged when it gains or loses negative charges.

      1. Once it moves, it stays put.

      2. It only jumps if it has a path.

   B. Current electricity is a steady flow of moving charges.

      1. An electric current follows an electric circuit.

      2. A battery provides the energy to move.

   C. Current electricity is more useful than static electricity.

II. Series Circuits

   _____

   _____

III. Parallel Circuits

   _____

   _____

IV. Resistance

   _____

   _____

Science Content Support   Use with Unit 1.

Name _____
Date _____

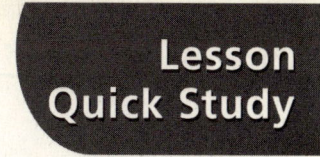

Lesson Quick Study

Unit 1, Lesson 2

Lesson 2—What Makes a Circuit?

1. **Investigation Skill Practice – Predict**

   Ben is wiring a circuit for his school diorama. He designs a parallel circuit for the lights because he wants to have some lights turn on at different times. However, Ben is not being careful. One of his wires runs from one end of his battery to the other end. It is not connected to light a bulb. What do you predict will happen when he tries to turn some of his lights on?

   _____
   _____
   _____
   _____

2. **Focus Skill — Reading Skill Practice – Cause and Effect**

   Read the selection. Describe what causes a string of lights to go out. Then describe the effect of breaking one light in the series circuit.

   Lydia bought a string of lights with a series circuit. John did not see the string of lights on the floor. He stepped on the string and broke a light bulb. Oh no! All the lights in the string went off. Lydia knew that the lights went out because current flowing in the series circuit was interrupted. She found a spare light bulb and replaced the broken one. All the lights came on again. Lydia decided not to let her string of lights lay on the floor ever again.

   _____
   _____
   _____
   _____

Use with Unit 1.   (page 1 of 2)   Science Content Support CS 23

Name _____

Science Concepts

3. Look at these pictures. Write *Parallel Circuit* or *Series Circuit* next to the correct pictures. Then answer the questions below.

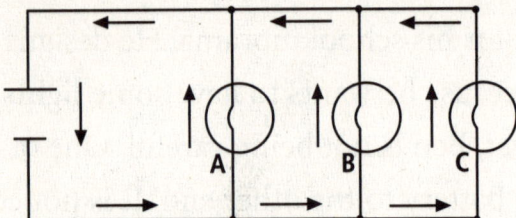

_____

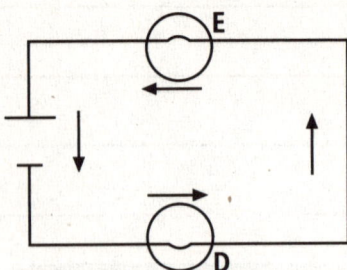

_____

4. What would happen if bulb B were removed?

_____

5. What would happen if bulb E were removed?

_____

6. Which type of circuit would you use if you were wiring the headlights and horn on a go-cart? Explain why.

_____

7. On the drawings above, draw a wire that would short circuit each electrical circuit.

_____

CS 24   Science Content Support

Lesson 2—What Makes a Circuit?

A. Simple Circuits

A circuit is a path that starts and stops at the same point. An electrical circuit is a circuit that allows electrical current to flow. A basic circuit includes batteries, bulbs, and wires. Circuits can also include switches, buzzers, motors, electromagnets, and other parts.

A circuit diagram shows how the parts of a circuit are organized. The symbols shown in the table represent the different parts of a circuit.

Circuit Part	Symbol
wire	——
battery	⊣⊢
bulb	⊗
switch	⟋
buzzer	⌓
motor	Ⓜ
electromagnet	⌇⌇⌇

B. Series Circuits

A series circuit provides only one path for current to flow.

Build the series circuit shown in this diagram.

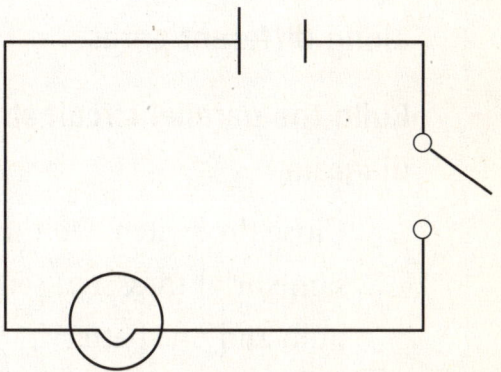

Close the switch. What happens?

_____

Open the switch. What happens?

_____

Name _____

Add a second bulb to the circuit.

Close the switch. Do both bulbs light?

_____

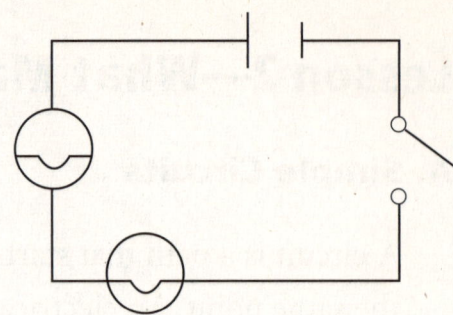

Compare the brightness of the bulbs to the circuit with only one bulb.

_____

Add a second battery to the circuit.

Close the switch. Compare the brightness of the bulbs to the circuit with two bulbs and one battery.

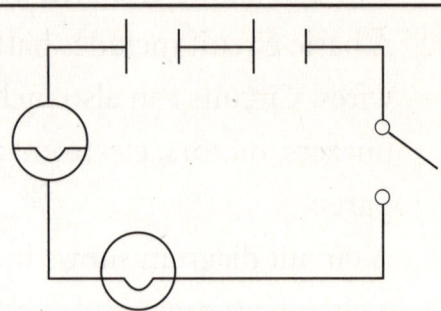

_____

Compare the brightness of the bulbs to the circuit with one bulb and one battery.

_____

C. Parallel Circuits

The current in a parallel circuit can travel along different paths.

Build the parallel circuit shown in this diagram.

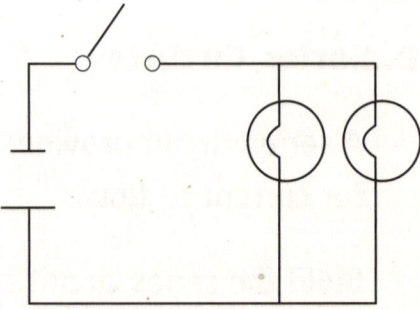

Close the switch. Do the bulbs light with same brightness as a series circuit with one bulb and one battery?

_____

Compare the brightness of two bulbs and one battery connected in parallel to two bulbs and one battery connected in series.

_____

Name _____

Add a third branch and bulb to the circuit.

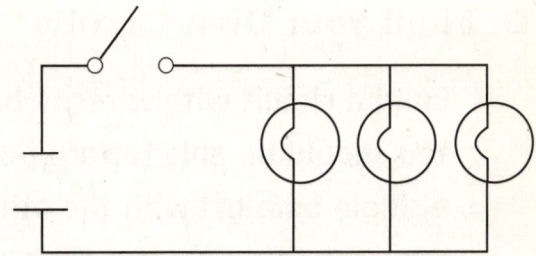

Close the switch. How does the brightness compare to the circuit with two bulbs?

_____

Move and add switches to create the circuit shown here.

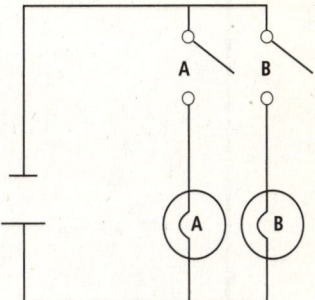

Close both switches. What happens?

_____

Close switch A and open switch B. What happens?

_____

Close switch B and open switch A. What happens?

_____

_____

Name _____

D. Build Your Own Circuits

Build a circuit with wires, 1 battery, 2 bulbs, and two switches. You should be able to turn both bulbs off with one switch and a single bulb off with the other switch.

Build a parallel circuit with any combination of wires, batteries, bulbs, and switches so that one bulb can be switched on and off and the other is on all of the time. Draw a circuit diagram to show your design.

Name _____

Date _____

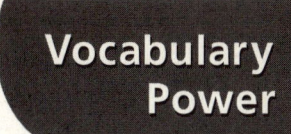

Unit 1, Lesson 3

Lesson 3—What Are Magnetic Poles?

A. Explore Word Meanings

Match the word on the left with the definition on the right.

1. _____ south-seeking pole
2. _____ magnetic poles
3. _____ magnet
4. _____ north-seeking pole
5. _____ magnetic field

a. an object that attracts iron and some other metals

b. the places on a magnet where the force is the strongest

c. the space around a magnet in which the magnetic force acts

d. the end of a magnet that will always point north

e. the end of a magnet that will always point south

B. Context Clues

Use the words from the box to fill in the blanks.

| magnets | north-seeking pole |
| magnetic poles | south-seeking pole |

Teddy learned about _____ from his train set. Each train car had a magnet on the front and the back. The magnets were used to link the cars together. The train cars would stay linked only when the opposite _____ faced each other. Each train car had one end with a _____. The other end had a south-seeking pole. When the north-seeking pole was facing the _____ the train cars would stay together. When they were not lined up, Teddy turned one of the cars around until the magnets attracted each other.

Use with Unit 1. Science Content Support CS 29

Name _____
Date _____

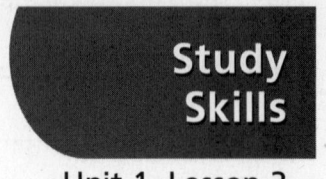

Study Skills

Unit 1, Lesson 3

Lesson 3—What Are Magnetic Poles?

Take Notes

Taking notes can help you remember important ideas.
- Write down important facts and ideas. Use your own words. You do not have to write in complete sentences.
- One way to organize notes is in a chart. Write down the main ideas in one column and facts and details in another.

As you read this lesson, use the two-column chart below to take notes.

What Are Magnetic Poles?	
Main Ideas	**Facts**
• A magnet has north-seeking and south-seeking poles.	• A magnet will attract one side of another magnet and will repel the other side.
• _____	• _____
• _____	• _____
• _____	• _____
• _____	• _____

CS 30   Science Content Support   Use with Unit 1.

Name _____
Date _____

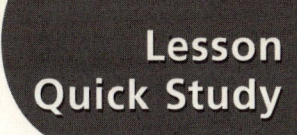

Lesson Quick Study

Unit 1, Lesson 3

Lesson 3—What Are Magnetic Poles?

1. **Investigation Skill Practice–Predict**

    Henri was in charge of making handbags at the factory. Each bag was supposed to have two magnets at the top to keep it closed. But Henri forgot to tell his workers to sew the magnets so that the opposite poles faced one another. Can you predict what problem Henri had when he got a shipment of handbags from his factory?

    _____
    _____
    _____
    _____
    _____

2. **Reading Skill Practice–Main Idea and Details**

    Read the selection. Underline the main idea. List 3 details.

    It is useful to know that magnets do not attract all metal objects. For example, a magnet will not attract a soda can because cans are made of aluminum. Pennies are made of copper and zinc. They are not attracted to a magnet either. However, most pins are made with some iron. You can use a magnet to pick them up if a pin box spills on the floor.

    _____
    _____
    _____
    _____
    _____
    _____

Use with Unit 1.   (page 1 of 2)   Science Content Support   CS 31

Science Concepts

3. **The statements below have some mistakes. Correct each statement on the line.**

   Each magnet has two poles: the east-seeking and west-seeking poles.
   _____
   _____

   The opposite poles on two different magnets repel one another.
   _____
   _____

   If you find that two magnets on a latch repel one another, you should throw one away because it must be broken.
   _____
   _____

   If you cut one magnet in half, one would be completely north-seeking and the other would be completely south-seeking.
   _____
   _____

   A magnetic field around a magnet is always round.
   _____
   _____

Name _____
Date _____

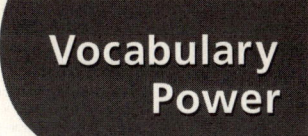

Unit 1, Lesson 4

Lesson 4—How Can You Detect a Magnetic Field?

A. Classify/Categorize

Circle the word in the list that does not belong.

1. sails   compass   lifeboats   chimney

2. sandstone   saw   lodestone   granite

B. Analogies

Fill in the missing part of the analogies with a word from the box.

| north-seeking | lodestone | compass |

3. *black* is to *white* as *south-seeking* is to _____

4. _____ is to *magnetic* as *glass* is to *see-through*

5. *thermometer* is to *temperature* as _____ is to *direction*

C. Explore Word Meanings

Write three sentences using each of the words from the box.

| lodestone | north-seeking pole | compass |

_____

_____

_____

Use with Unit 1.   Science Content Support   CS 33

Name _____
Date _____

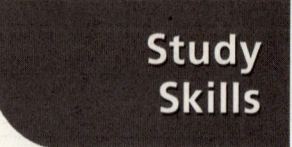

Unit 1, Lesson 4

Lesson 4—How Can You Detect a Magnetic Field?

Pose Questions

Posing or asking questions as you read can help you understand what you are reading.

- Form questions as you read. For example, you may ask how a science concept is connected to other concepts.
- Use the questions to guide your reading. Look for answers as you read.

Before you read this lesson, write a list of questions in the chart below. Look for the answers as you read. Record the answers in the chart.

How Can You Detect a Magnetic Field?	
Questions	Answers
Why does a compass sometimes point in the wrong direction?	Compasses can be influenced by the magnetic field of other objects.

CS 34   Science Content Support

Use with Unit 1.

Name _____
Date _____

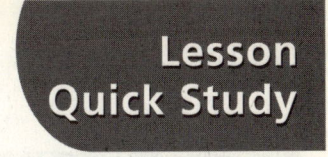

Unit 1, Lesson 4

Lesson 4—How Can You Detect a Magnetic Field?

1. **Investigation Skill Practice – Predict**

   Jeremy holds his compass as he walks through his house. Which of these items do you predict will make the compass' needle point away from the magnetic north pole? Write *yes* or *no* next to each item.

   _____ his computer   _____ his sandbox

   _____ his beanbag chair   _____ the car

   _____ the oven   _____ plastic toys

   _____ the laundry basket   _____ the television

   _____ a tree in his yard

2. **Reading Skill Practice – Cause and Effect**

   Read the selection. Explain what caused the sailors to get lost before the compass was invented. Describe the effect of the compass on sailing.

   Early sailors knew their direction from the position of the sun. But sometimes, ships sailed for days without seeing the sun. Sometimes sailors would get lost. They could not tell their direction on cloudy days. Then the compass was invented. The compass meant that even on dark stormy days, the sailors knew which direction they were sailing. With a compass, sailors did not have to stay within sight of land. They could explore much further from their own shore. Soon explorers were sailing all over the world. They learned about huge oceans and about great masses of land.

   _____
   _____
   _____
   _____

Use with Unit 1.   (page 1 of 2)   Science Content Support   CS 35

Name _____

Science Concepts

3. Write *True* or *False* in the blanks.

   _____ **A.** The magnetic North Pole is the same as the geographic North Pole.

   _____ **B.** The Earth's core acts as a magnet.

   _____ **C.** A compass is only useful for sailors on cloudy days.

   _____ **D.** Most compasses are small, lightweight, and easy to use.

   _____ **E.** The best place to use a compass is outside and away from metal buildings or objects.

   _____ **F.** Wooden and plastic objects will create magnetic interference on your compass.

4. Answer the questions below.

   A. Why would the steel in a modern ship affect a compass?

   _____

   _____

   B. Write a plan to figure out how can you be sure you are not getting magnetic interference with your compass.

   _____

   _____

   C. How is the magnetic field around a bar magnet similar to Earth's magnetic field?

   _____

   _____

Name _____
Date _____

Unit 1, Lesson 5

Lesson 5—What Makes an Electromagnet?

A. Using Prefixes

The prefix *electro-* means "the use of electricity in." Choose a word from the box for each definition in the chart.

| electrochemistry | electromagnet | electrometallurgy |

Word	Definition
_____	the use of electricity to produce a chemical reaction
_____	the branch of metallurgy that uses electricity
_____	a magnet made of iron wrapped with a coil of current-carrying wire

B. Explore Word Meanings

Complete each sentence.

1. An **electromagnet** can be described as
_____

2. An **electromagnet** is different from a magnet because
_____

3. An **electromagnet** is useful because
_____

Use with Unit 1.   Science Content Support   CS 37

Name _____

Date _____

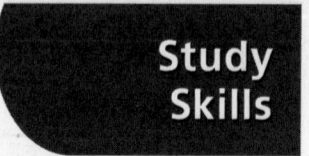

Unit 1, Lesson 5

Lesson 5—What Makes an Electromagnet?

Organize Information

Graphic organizers can help you organize information and make sense of the facts you read.

- Tables, charts, and webs are graphic organizers that can show main ideas and important details.
- Graphic organizers help you categorize, or group, information.
- Putting related ideas into categories makes it easier to find facts.

As you read the lesson, fill in this graphic organizer about electromagnets.

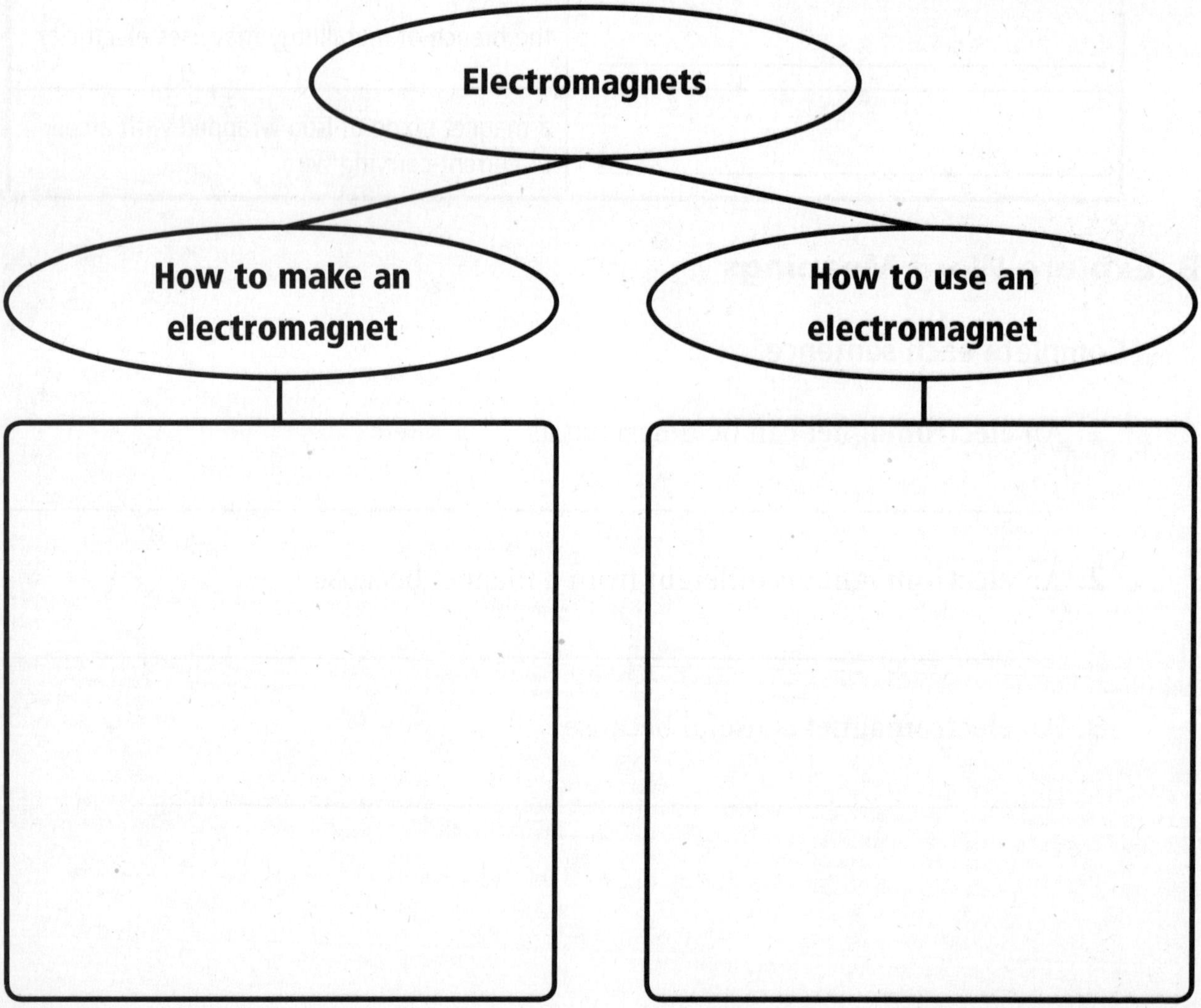

Electromagnets
├── How to make an electromagnet
└── How to use an electromagnet

CS 38 Science Content Support

Use with Unit 1.

Name _____

Date _____

Lesson Quick Study

Unit 1, Lesson 5

Lesson 5—What Makes an Electromagnet?

1. Investigation Skill Practice—Making and Interpreting Graphs

Janet made this graph when she made her electromagnet. Look at the chart and answer the questions that follow.

Experiment Number	Number of Coils of Wire	Number of Clips Lifted
1	1	0
2	10	3
3	20	4
4	50	17
5	100	38

What happened to the number of clips lifted as Janet added coils of wire?

_____

Why do you think Janet's electromagnet did not lift any clips on her first experiment?

_____

Make a line graph that shows Janet's experiment. Use the graph below to chart the number of coils of wire and the number of clips lifted.

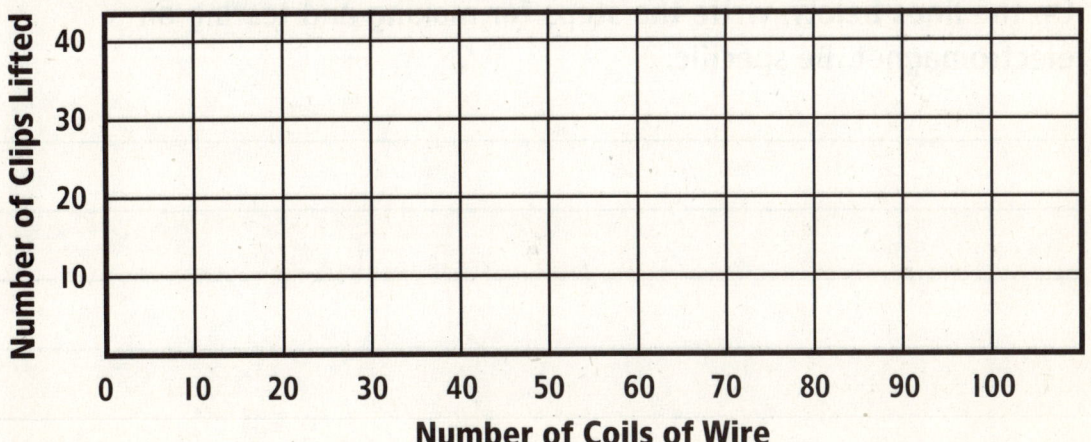

Use with Unit 1. (page 1 of 2) Science Content Support CS 39

Name _____

2. **Reading Skill Practice—Main Idea and Details**

   Read the selection. Underline the main idea. Circle two details.

   An electromagnet may seem like a useless toy, but it can be very useful. At a construction site, a big machine raises and lowers large steel beams. It uses a powerful electromagnet. To pick up the steel beam, the operator turns the magnet on. The machine lifts the beam high up on the building where it is needed. To release the beam, the operator turns the electromagnet off. It is ready to lift the next beam.

Science Concepts

3. Draw the magnetic field around the magnets below. Draw lines close together where the magnetic field is strong. Draw lines far apart where the magnetic field is weak.

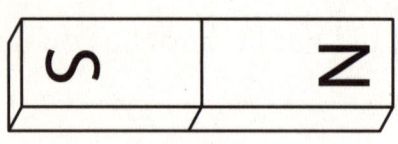
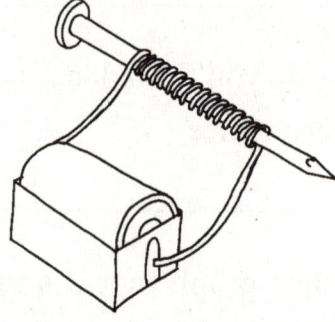

4. On the lines below, write the steps for making and testing an electromagnet. Be specific.

   _____
   _____
   _____
   _____

Name _____

Date _____

Vocabulary Power

Unit 1, Lesson 6

Lesson 6—How Are Electromagnets Used?

A. Explore Word Meanings

Match the vocabulary word on the left with the definition on the right.

1. _____ generator
2. _____ kinetic energy
3. _____ relay
4. _____ electric motor
5. _____ alternating current

a. electricity that is always changing direction back and forth

b. a device that converts electrical energy to kinetic energy

c. a device that converts other forms of energy into electrical energy

d. a device that uses electricity in one circuit to close a switch in a different circuit

e. the energy of motion

B. Context Clues

Fill in the blanks with words from the box.

| generator | kinetic energy | electric motor |

FLASH! A big bolt of lightning struck. The lights went out. The house went black. Ian and Edward blinked their eyes. Everything in the house with an _____ had shut down. They thought about what to do. They needed Dad's _____ powered radio. They felt along the walls to the basement and found it in a corner. They were lucky that the machine used kid-powered _____ to operate. They worked the crank and soon the radio was playing weather news.

Use with Unit 1.

Science Content Support CS 41

Name _____

Date _____

Study Skills

Unit 1, Lesson 6

Lesson 6—How Are Electromagnets Used?

Preview and Question

Identifying main ideas in a lesson and asking question about them can help you find important information.

- To preview a lesson, read the lesson title and the section titles. Look at the pictures, and read their captions. Try to get an idea of the main topic and think of questions you have about the topic.
- Read to find the answers to your questions. Then recite, or say, the answers aloud. Finally, review what you have read.

As you read this lesson, fill in the chart and practice reading, reciting, and reviewing.

How Are Electromagnets Used?				
Preview	Questions	Read	Recite	Review
Build a Motor: Motors contain electromagnets	How can an electromagnet make something move?	✔	✔	✔

CS 42 Science Content Support

Use with Unit 1.

Name _____
Date _____

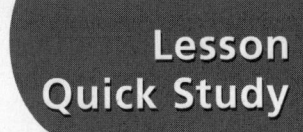

Unit 1, Lesson 6

Lesson 6—How Are Electromagnets Used?

1. Investigation Skill Practice—Following Written Directions

Brenda got a new electric set for her birthday. She figured she already knew how to make a motor, so she did not read the instructions. Instead she put the wires and magnets together the way she remembered. Then she plugged in her motor. POW! She got a bright flash and a smoky smell. Her motor did not work, and her wires were burned up. What do you think Brenda could have done to avoid this problem? Explain your answer.

_____
_____
_____
_____

2. Reading Focus Skill Practice—Compare and Contrast

Read the selection. Compare and contrast the two radios.

Rashid needed a radio for his camping trip. He went shopping for radios that did not need to be plugged in. He found two radios that were about the same price. The first radio used four D batteries. The batteries let the radio play for 12 hours before they needed to be replaced. The second radio was powered by a hand crank. He had to crank the radio for one minute to make it play for two hours. Rashid wondered which one was the best buy.

_____
_____
_____
_____

Use with Unit 1.   (page 1 of 2)   Science Content Support   CS 43

Name _____

Science Concepts

3. **Powering a Generator**

   Fill in the flow charts below with four types of energy that could power a generator to produce electricity.

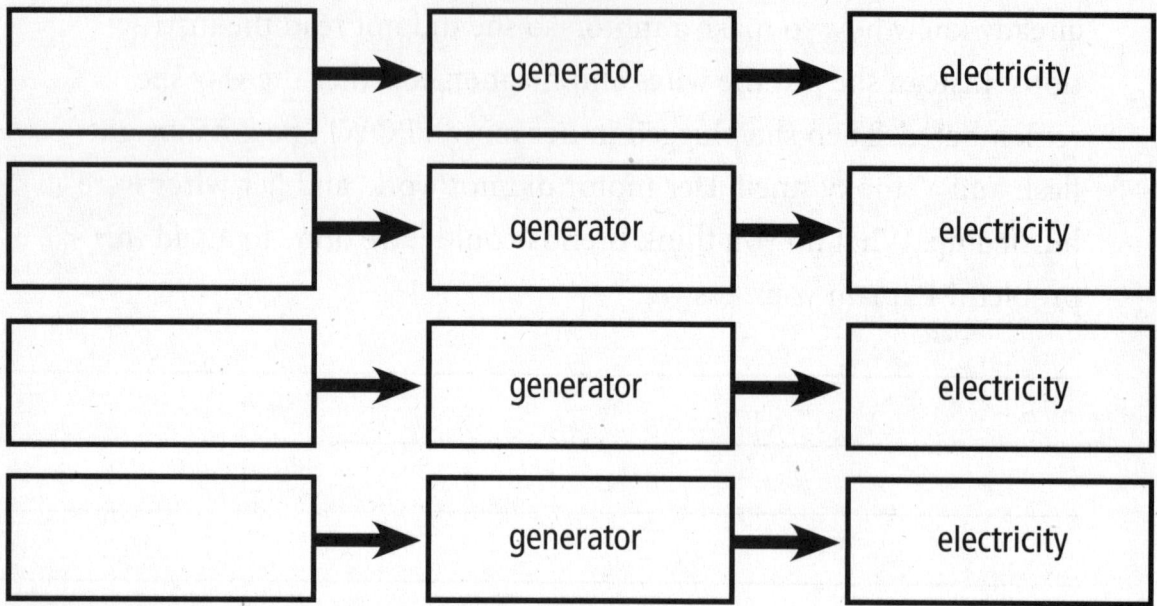

4. **Using a Motor**

   Read the steps in the box. Write one step in each box of the flow chart to show the process of making a fan.

   ```
   electric energy goes to motor
   motor turns blades of fan
   permanent magnet pushes and pulls shaft of motor
   wire coils become electromagnet
   ```

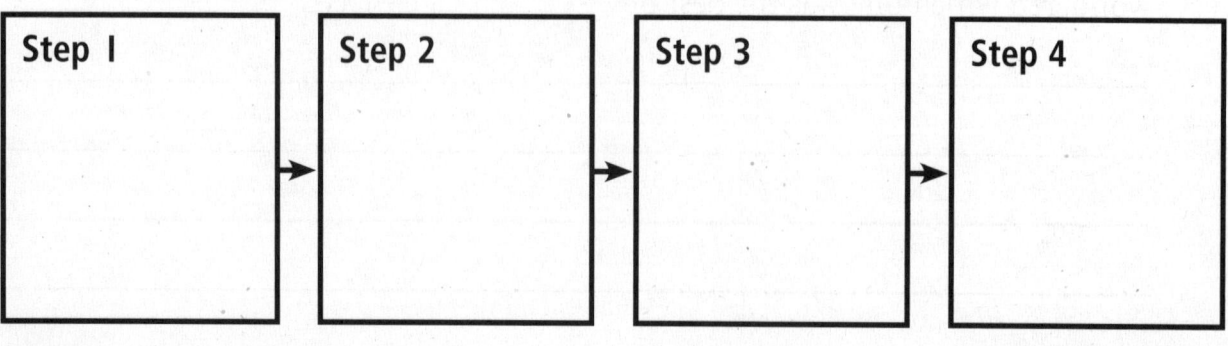

Use with Unit 1.

Name _____
Date _____

Unit 1, Lesson 7

Lesson 7—How Is Electrical Energy Used?

A. Word Families

Some words can take on different forms. Read the example. Then fill in as many forms of the following words that you can think of.

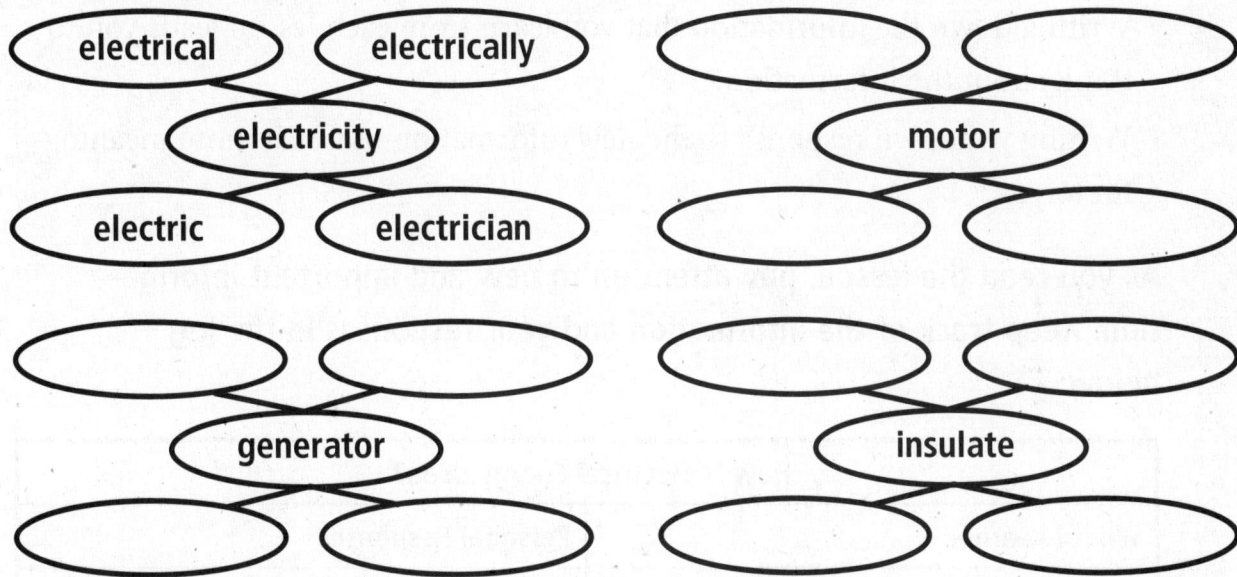

B. Context Clues

Fill in the blanks in the paragraph with a word from the box.

| insulate | safety | shock | electricity |

When Theodore's electric drill stopped working, he was disappointed. He did not have the money to buy a new drill. He unplugged it. Then his dad took off the cover and looked at the wiring. He noticed that the wires from the cord were frayed. The rubber used to _____ the wires had gotten old and broken. Theodore was worried about electrical _____. He did not want to get a _____ when he used his drill. His father decided to repair the frayed wires. He carefully reconnected them to the motor. The wires now carried _____. His drill was fixed.

Use with Unit 1.   Science Content Support   CS 45

Name _____
Date _____

Study Skills

Unit 1, Lesson 7

Lesson 7—How Is Electrical Energy Used?

Write to Learn

Writing about what you read can help you better understand and remember information.

- Writing down the information that you learn from each lesson leads you to think about the information.
- Writing your own response to the new information makes it more meaningful to you.

As you read the lesson, pay attention to new and important information. Keep track of the information and your responses in the log below.

How Is Electrical Energy Used?	
What I Learned	**Personal Response**
We depend on electricity for heat and light.	Imagining living without electricity makes me realize how much our lives depend on it for almost every activity.

CS 46 Science Content Support   Use with Unit 1.

Name _____

Date _____

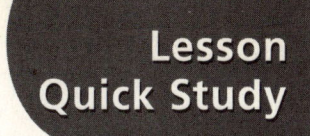

Unit 1, Lesson 7

Lesson 7—How Is Electrical Energy Used?

1. **Investigation Skill Practice–Infer**

   Read the paragraph. Answer the questions that follow.

   When Teresa moved into an old apartment building, she noticed something strange. She noticed that a circuit breaker would go off in her house whenever the hairdryer and the toaster were used at the same time. She tried an experiment. She turned on the toaster and then tried other electrical devices at the same time. She tried using the blender, the radio, and the lamp. None of these appliances made the fuse blow.

   What can you infer about the wiring in the old apartment building?

   _____

   What can you infer about the amount of electricity required by the toaster and the hairdryer?

   _____

   What can you infer about the amount of energy electric heating devices use?

   _____

2. **Reading Focus Skill Practice–Main Idea and Details**

   Read the selection. Underline the main idea. Circle three details.

   Using a power strip is a good way to use electricity safely. A power strip plugs into one wall socket. You can plug many appliances into the same power strip. Power strips use thick wires. They do not heat up when many appliances are plugged in. Power strips are made to shut off if the electricity surges in your house.

Name _____

Science Concepts

3. Read the list of devices. Write a check mark for each device under the column if it Produces Heat, Produces Light, or Produces Motion. You may check more than one column. In the last column, tell what you would use to replace that item if you did not have electricity.

Device	Produces Heat	Produces Light	Produces Motion	Non-electric Device to Replace It
lamp				
computer				
air conditioner				
fan				
mixer				
blender				
toaster				
oven				
microwave				
television				

4. Write down three rules for handling electricity safely.

_____
_____
_____
_____

CS 48   Science Content Support   (page 2 of 2)   Use with Unit 1.

Name: _____
Date: _____

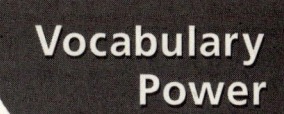

Unit 2, Lesson 1

Lesson 1—What Are Producers and Consumers?

Context Clues

Study the words and their definitions below.

> **carnivore**– an animal that eats only other animals
> **consumer**– a living thing that can't make its own food and must eat other living things
> **herbivore**– an animal that eats only plants or producers
> **omnivore**– an animal that eats both plants and other animals
> **photosynthesis**– the process in which water and carbon dioxide are combined in the presence of sunlight to form sugars and oxygen
> **producer**– a living thing, such as a plant, that can make its own food

Use the underlined context clues to help fill in the blanks with one of the words above. Use all of the words once.

1. An organism, such as a bush or a tree, that makes it own food, is a _____.

2. By eating both plants and other animals, people are considered _____.

3. A _____ is an organism that only consumes the flesh of other animals.

4. An organism that cannot make its own food is a _____.

5. Some vegetarians can be called _____ because they avoid all meat and eat only plants.

6. A plant makes its own food using water, sunlight, and carbon dioxide in a process called _____.

Use with Unit 2. Science Content Support CS 49

Name _____
Date _____

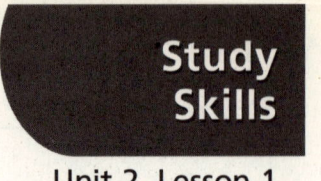

Unit 2, Lesson 1

Organize Information

Graphic organizers can help you organize information and make sense of the facts you read.

- Tables, charts, and webs are graphic organizers that can show main ideas and important details.
- Graphic organizers help you categorize, or group, information.
- Putting related ideas into categories makes it easier to find facts.

As you read the lesson, fill in the graphic organizer below about living organisms.

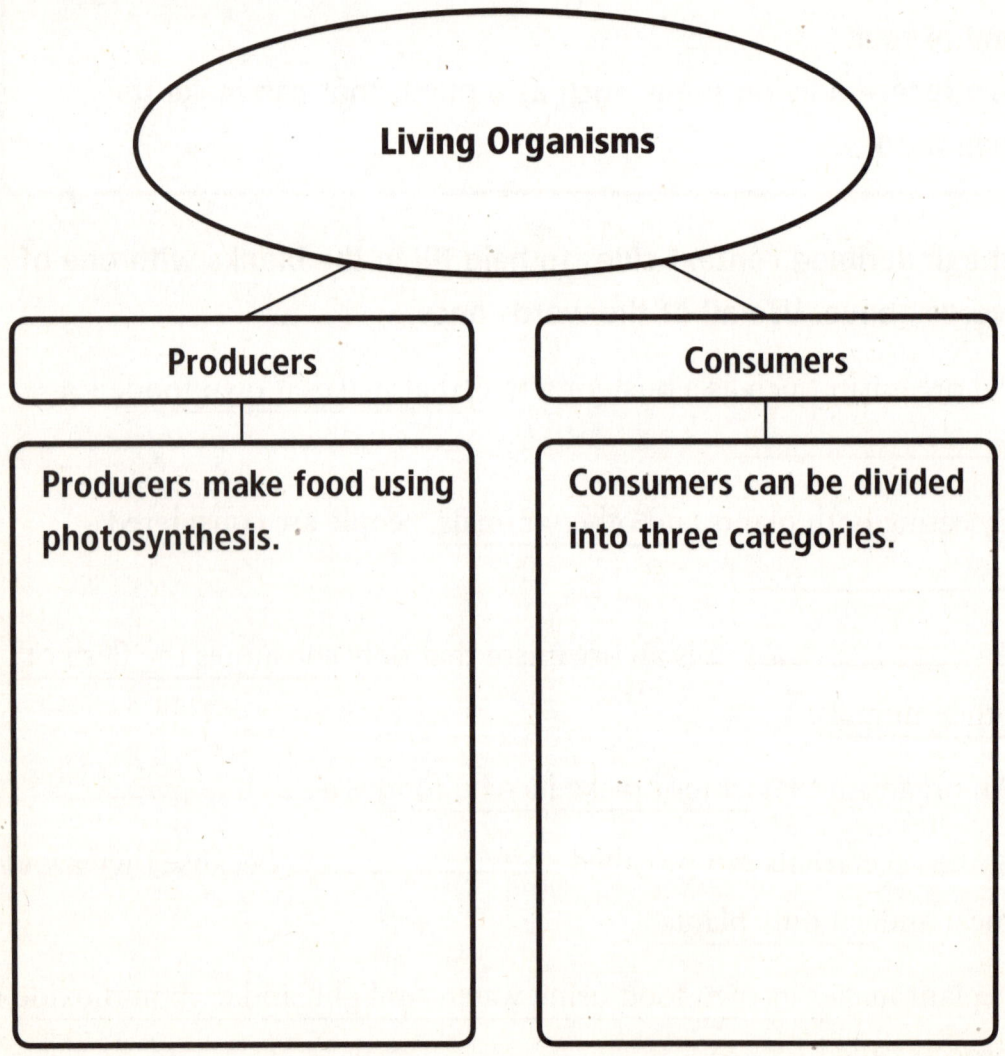

CS 50   Science Content Support

Use with Unit 2.

Name _____

Date _____

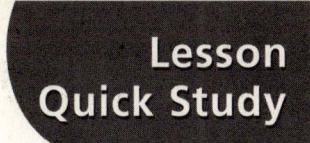

Lesson Quick Study

Unit 2, Lesson 1

Lesson 1—What Are Producers and Consumers?

1. **Investigation Skill Practice–Classify**

   Classify each organism in the box below as a producer or a consumer. Write the name of each organism in the *producer* or *consumer* column in the chart below.

redwood tree	grape vine	penguins	tomatoes
algae	grass	bush	walrus
eagle	hyena	shrimp	
elephants	moss	spiders	

Producers	Consumers

2. **Reading Skill Practice–Main Idea and Details**

   Read the selection. Underline the main idea. List the details on the lines below.

   Some birds are carnivores, such as eagles, hawks, and owls. However, some birds you see at your bird feeder are omnivores. In the summer time, these wild birds eat like carnivores, munching on caterpillars, worms, and insects. They do not rely on insects and worms to survive the winter! Wild birds eat berries on bushes. They will even pick the seeds out of pinecones to fill their bellies when the snow flies.

   _____

   _____

Use with Unit 2.   (page 1 of 2)   Science Content Support   **CS 51**

Name _____

Science Concepts

3. Some organisms need sunlight, water, and carbon dioxide to survive. Other organisms need much more. Choose the type of food from the box that each organism needs to survive.

Types of Food	
carbon dioxide	sunlight
meat	water
plants	

A. grass

_____

B. deer

_____

C. lions

_____

D. bears

_____

E. raccoons

_____

F. pine trees

_____

4. Answer the questions using the organisms above.

A. Which organisms are producers?

_____

B. Which organisms are carnivores?

_____

C. Which organisms are herbivores?

_____

D. Which organisms are omnivores?

_____

Name _____

Date _____

Unit 2, Lesson 2

Lesson 2—What Are Decomposers?

Context Clues and Explore Word Meanings

Read each sentence or sentences. Use context clues to find the definition of each word in the chart below. Write the word next to the correct definition in the chart.

1. As the wagon train rode west, an ox dropped to the ground, exhausted and dying. It was left to the **scavengers**, who hovered around, eating the flesh of the dead animal.

2. After the buzzards had picked the bones clean, tiny **microorganisms** set to work on the scraps.

3. Later the spores from a **fungus** were blown by the wind to the rich soil under the bones. The spores grew quickly into mushrooms.

4. A year later, many **decomposers** had finished breaking down the body of the ox and nothing but the bones remained in the hot sun.

	a living thing that feeds on the wastes of plants and animals
	a living thing that feeds on dead organisms
	an organism that can't make food and can't move about
	an organism that is too small to be seen without a microscope

Use with Unit 2.   Science Content Support   CS 53

Name _____
Date _____

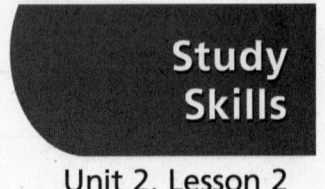

Study Skills

Unit 2, Lesson 2

Lesson 2—What Are Decomposers?

Pose Questions

Posing or asking questions as you read can help you understand what you are reading.

- Form questions as you read. For example, you may ask how a science concept is connected to other concepts.
- Use the questions to guide you reading. Look for answers as you read.

Before you read this lesson, write a list of questions in the chart below. Look for the answers as you read. Record the answers in the chart.

What Are Decomposers?	
Questions	Answers
What kinds of animals are considered decomposers?	Decomposers are bacteria, scavengers, fungi, and microorganisms.
• _____ • _____	• _____ • _____
• _____ • _____	• _____ • _____
• _____ • _____	• _____ • _____

Science Content Support

Use with Unit 2.

Name _____
Date _____

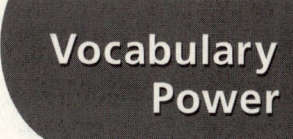

Unit 2, Lesson 2

Lesson 2—What Are Decomposers?

Context Clues and Explore Word Meanings

Read each sentence or sentences. Use context clues to find the definition of each word in the chart below. Write the word next to the correct definition in the chart.

1. As the wagon train rode west, an ox dropped to the ground, exhausted and dying. It was left to the **scavengers**, who hovered around, eating the flesh of the dead animal.

2. After the buzzards had picked the bones clean, tiny **microorganisms** set to work on the scraps.

3. Later the spores from a **fungus** were blown by the wind to the rich soil under the bones. The spores grew quickly into mushrooms.

4. A year later, many **decomposers** had finished breaking down the body of the ox and nothing but the bones remained in the hot sun.

	a living thing that feeds on the wastes of plants and animals
	a living thing that feeds on dead organisms
	an organism that can't make food and can't move about
	an organism that is too small to be seen without a microscope

Use with Unit 2. Science Content Support CS 53

Name _____
Date _____

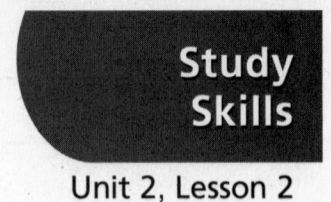

Study Skills

Unit 2, Lesson 2

Lesson 2—What Are Decomposers?

Pose Questions

Posing or asking questions as you read can help you understand what you are reading.

- Form questions as you read. For example, you may ask how a science concept is connected to other concepts.
- Use the questions to guide you reading. Look for answers as you read.

Before you read this lesson, write a list of questions in the chart below. Look for the answers as you read. Record the answers in the chart.

What Are Decomposers?	
Questions	Answers
What kinds of animals are considered decomposers?	Decomposers are bacteria, scavengers, fungi, and microorganisms.
• _____ • _____	• _____ • _____
• _____ • _____	• _____ • _____
• _____ • _____	• _____ • _____

CS 54 Science Content Support

Use with Unit 2.

Name _____
Date _____

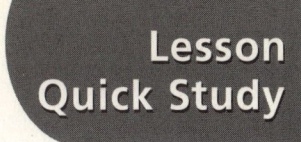

Lesson Quick Study

Unit 2, Lesson 2

Lesson 2—What Are Decomposers?

1. **Investigation Skill Practice–Use Time Relationships**

   Read the list of what happens to the body of a dead animal over time.

Day 1: A California condor eats chunks of the dead animal.
Day 2: Crows pick at bones of the dead animal.
Day 3: Flies lay eggs in the meat of the dead animal.
Day 4: Maggots hatch and eat the flesh.
Day 5: Bacteria grow in the animal carcass.

   On the lines below, describe what happens to the dead animal over time.

   _____
   _____
   _____

2. **Reading Focus Skill Practice–Compare and Contrast**

   Using the selection below, explain how the mushrooms are alike. Explain how they are different.

   Both morel and Portobello mushrooms can be found on the menus at fancy restaurants. They both have a meaty texture and improve the flavors of sauces and seasonings. However, the two mushrooms are very different from one another. The morel is shaped like a sack. It has a rough textured skin. It is usually dried before it is cooked. The Portobello is shaped like a classic mushroom with a stem and a smooth cap. Portobello mushrooms are easy to find fresh in most grocery stores.

   _____
   _____
   _____

Name _____

Science Concepts

3. For each decomposer listed in the left column, check (✔) the box that describes how it lives and eats.

Decomposers	Eats animal flesh	Eats dead plants	Lays eggs in animal flesh	Returns nutrients to the environment	Sources of nutrients for other creatures	Decomposes wood
mold						
yeast						
bacteria						
mushroom						
fly						
pilobolus						
bracket fungus						
sac fungus						
slime mold						

4. On the lines below, summarize the role of a decomposer in breaking down other organisms.

_____

_____

_____

_____

_____

Name _____

Date _____

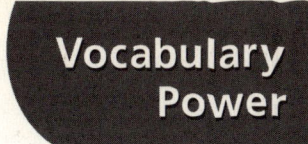

Unit 2, Lesson 3

Lesson 3—What Are Food Chains and Food Webs?

A. Context Clues

Read the sentences and use the context clues to choose the correct meaning of the underlined word. Circle its meaning. Look up the word in a glossary if you need help.

1. The deer became <u>prey</u> to the mountain lion that attacked at dawn.
   consumer eaten by a predator
   consumer eaten by a producer

2. The top <u>predator</u> in the jungle was the eagle, whose sharp eyes caught the movement of the smallest monkey.
   consumer of plants
   consumer of other animals
   consumer of dead organisms

3. When the wolves died and stopped eating the rabbits, the <u>food chain</u> was disrupted and rabbits began to overrun the prairie.
   group of animals that live in the same hole
   series of organisms that depend on one another for food
   group of animals that all have the same mother

4. Although the blue whale is the largest animal on earth, it does not eat other fish. Its diet consists only of tiny <u>plankton</u>.
   small producers in the ocean
   small pine needles
   fish and seals

5. The food chain starts with a <u>producer</u> that makes food from sunshine, water, and carbon dioxide.
   first source of shelter for living things
   first source of water for living things
   first source of energy for living things

Use with Unit 2.   Science Content Support   CS 57

Name _____
Date _____

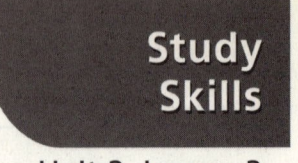

Unit 2, Lesson 3

Lesson 3—What Are Food Chains and Food Webs?

Use Visuals

Visuals can help you better understand and remember what you read.

- Photographs, illustrations, diagrams, charts, and maps are different kinds of visuals.
- Many visuals have titles, captions, or labels that help readers understand what is shown.
- Visuals often show information that appears in the text, but they show it in a different way.

As you read this lesson, look closely at the visuals and the text that goes with them. Answer the questions in the checklist.

	What Are Food Chains and Food Webs?
✓	What kind of visual is shown to illustrate a food chain and a food web? _____
✓	What does the visual show? _____
✓	How does the visual relate to the lesson you are reading? _____
✓	How does the visual help you better understand the subject of what you are reading? _____

Science Content Support

Use with Unit 2.

Name _____

Date _____

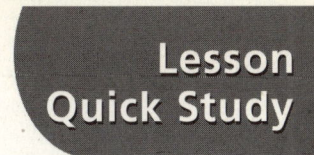

Lesson Quick Study

Unit 2, Lesson 3

Lesson 3—What Are Food Chains and Food Webs?

1. **Investigation Skill Practice—Communicate**

   Look at the list of organisms from the desert. Think about how they are related in a food web. Then write a brief paragraph to communicate how the organisms might depend on one another.

bats	sagebrush	hawks	snakes
birds	foxes	insects	

   _____

   _____

   _____

   _____

2. **Reading Skill Practice—Sequence**

   Put the following events about food chains and food webs in the correct sequence. Number the steps 1 to 5.

   _____ Bacteria breaks down the remains of a dead hawk.

   _____ Plants grow in the lush forest.

   _____ A field mouse feeds on grass seeds.

   _____ A hawk carries a weasel away for its next meal.

   _____ A weasel catches a mouse and eats it.

Use with Unit 2.   (page 1 of 2)   Science Content Support   CS 59

Name _____

Science Concepts

3. Organize the list of organisms below into food chains that include a producer, a first-level consumer, a second-level consumer, and a top-level consumer. Write the organisms on the chart. Some organisms belong to more than one food chain.

acorns	foxes	plankton	snails
algae	grass	polar bears	snakes
chipmunks	hawks	rabbits	wolves
fish	herons	seals	

	Food Chain 1	Food Chain 2	Food Chain 3	Food Chain 4
producers				
first-level consumers				
second-level consumers				
top-level consumers				

4. Explain how a food chain is different from a food web.

_____
_____
_____
_____

Name _____
Date _____

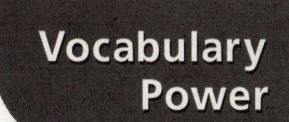

Unit 2, Lesson 4

Lesson 4—How Do Living Things Compete for Resources?

A. Root Words

Write the word from the box that uses the root word below. Write a definition for each word from the box.

| habitat | resources | competition |

1. source: the point where something begins

   _____
   _____

2. compete: to work or fight for something

   _____
   _____

3. habit: pattern of behavior

   _____
   _____

B. Explore Word Meanings

Think about the meaning of the underlined words. Then write your answer to each question.

4. How would you describe a desert <u>habitat</u>?

   _____
   _____

5. How would you describe an organism's <u>niche</u> in its habitat?

   _____
   _____

Use with Unit 2. Science Content Support CS 61

Name _____

Date _____

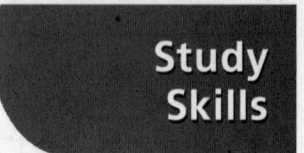

Study Skills

Unit 2, Lesson 4

Lesson 4—How Do Living Things Compete for Resources?

Skim and Scan

Skimming and scanning are two ways to learn from what you read.

- To skim, quickly read the lesson title and the section titles. Look at the visuals, or images, and read the captions. Use this information to identify the main topics.
- To scan, look quickly through the text for specific details, such as key words or facts.

Before you read this lesson, skim the text to find the main ideas. Then look for key words. If you have questions about a topic, scan the text to find the answers. Fill in the chart below as you skim and scan.

How Do Living Things Compete for Resources?	
Skim	**Scan**
Lesson Title:	Key Words and Facts:
Main Idea:	
Section Titles:	
Visuals:	

CS 62 Science Content Support

Use with Unit 2.

Name _____
Date _____

Lesson Quick Study

Unit 2, Lesson 4

Lesson 4—How Do Living Things Compete for Resources?

1. Investigation Skill Practice–Predict

One California city is growing too big for its water supply. Some people suggest that the city could reroute a river to make it flow closer to the city. Consider the habitat of the area where the river now flows. How do you predict it will change if the river is rerouted?

_____
_____
_____

2. Reading Focus Skill Practice–Cause and Effect

Read the selection. Describe the cause of the dirty river. Explain the effects.

The river that ran through Jan's town was full of garbage. The fish died and few birds drank from the river anymore. Jan and her neighbors decided to change the river. They helped get new laws passed. They said, "No more dumping in the river!" They cleaned out the garbage. Three years later, Jan noticed river weeds growing along the river banks. Then she saw a family of turtles living among the weeds. She could tell that the river habitat was becoming healthy again.

_____
_____
_____
_____

Name _____

Science Concepts

3. **Identify Habitats**

   In the box below, what items are part of the habitats listed in the chart below? Items may be used more than once or may not be used at all.

TV hookup	shelter	heavy rainfall	sagebrush
interstate	insects	snakes	sunlight
highways	producers	frogs	bulldozers
water source	dry weather	birds	radio towers

Desert Habitat	Rainforest Habitat	Woodlands Habitat

4. Explain how competition for resources helps keep animal populations from getting too large.

   _____
   _____
   _____
   _____

CS 64 Science Content Support Use with Unit 2.

Name _____

Date _____

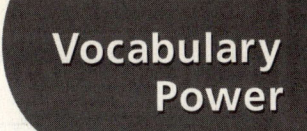

Unit 3, Lesson 1

Lesson 1—What Makes Up an Ecosystem?

A. Context Clues

Read each sentence. Predict the meaning of the underlined word from the context clues. Write the word next to the correct definition in the chart below.

1. The student studied the ecosystem by collecting samples of plants, recording numbers of animals, and keeping notes on rainfall in the area.

2. The scientist was worried about the population of bullfrogs in the pond because he noticed that many of them were born with strange deformities.

3. The jungle community supported the gorillas by supplying all they needed to live, such as plants to eat, trees for shelter, and scavengers to clean up after them.

4. The armadillo lives in a biome with a warm, dry climate and soil that supports the insects and plants that make up its diet.

5. By studying the ecology of a forest, the ecologist noticed that the birds needed to nest in the tallest trees to survive.

Word	Definition
	all the populations of organisms living in an environment
	a large area with a similar climate and ecosystem
	all the individuals of one kind living in the same environment
	the study of ecosystems
	a community of living things and its physical environment

Use with Unit 3.   Science Content Support   CS 65

Lesson 1—What Makes Up an Ecosystem?

Connect Ideas

You can use a web organizer to show how different ideas and information are related.

- List important themes in the ovals in the web's center.
- Add ovals showing main ideas that support each theme.
- Add bubbles for the details that support each main idea.

Complete the web as you read this lesson. Fill in each bubble by adding facts and details. Add more bubbles if you need them.

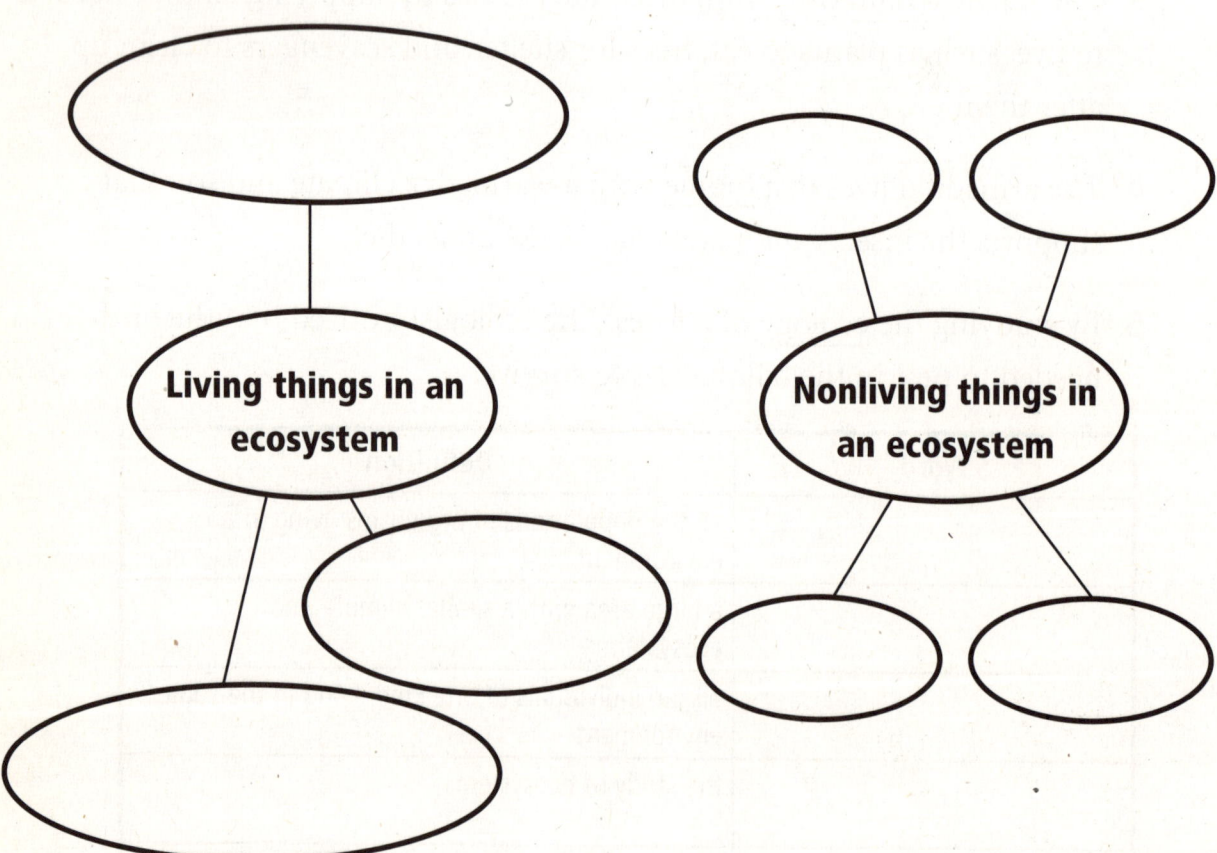

Name _____

Date _____

Lesson Quick Study

Unit 3, Lesson 1

Lesson 1—What Makes Up an Ecosystem?

1. Investigation Skill Practice–Using a Model

When you plant a garden, you create a mini ecosystem. It must survive within the ecosystem of your local climate. But it contains many of the elements of any other ecosystem around the world. On the lines below, use a vegetable garden as a model to explain how an ecosystem works.

_____
_____
_____
_____
_____

2. Reading Skill Practice–Main Idea and Details

Read the selection. Underline the main idea. Write 3 details on the lines below.

Some organisms can only survive in very specific ecosystems. Other organisms can adapt to almost any ecosystem. For example, the common cockroach is one of the most adaptable organisms on the planet. Cockroaches can survive in the woods and in the jungle. When humans cut down trees and drain swamps, the cockroach makes its home with people. It can live in cities and in towns. Cockroaches like barnyards and even apartment buildings.

_____
_____
_____
_____

Use with Unit 3.   (page 1 of 2)   Science Content Support   CS 67

Name _____

Science Concepts

3. Identify the Ecosystem

   Read each item or list of items below. Choose the word from the box that describes the item or group of items.

individual	population	community	ecosystem

   A. 30 red-headed woodpeckers
   _____

   B. seaweed, shellfish, coral, saltwater, sunlight, and fish of the ocean
   _____

   C. insects, spiders, pets, and humans living in my home
   _____

   D. soil, water, sunlight, plants, and animals of the taiga
   _____

   E. pet German shepherd
   _____

4. Answer the questions below.

   A. Give an example of different kinds of biomes.
   _____

   B. Describe the biome you live in.
   _____

Name _____
Date _____

Unit 3, Lesson 2

Lesson 2—What Affects Survival?

A. The Suffix –tion

The suffix –*tion* means "the act of doing something." Using the words below, write its root word on the line and make a prediction about the definition of the word.

Word	Root Word	Definition
1. hibernation	hibernate	the act of hibernating
2. absorption	_____	_____
3. prediction	_____	_____
4. investigation	_____	_____
5. reproduction	_____	_____
6. accommodation	_____	_____
7. adaptation	_____	_____

B. Complete the Sentence

Complete the sentences with the correct word from the list above.

1. One way that animals survive the cold winter is to go into a sleep-like state of _____.

2. Raccoons and bears are good at _____ because they learn new behaviors to help them survive when their environment changes.

3. Organisms have _____, such as thick fur, an ability to hibernate, or an ability to camouflage so they can survive.

C. Explore Word Meanings

Write two sentences on the lines below using the words *accommodation* and *adaptation* correctly.

_____

_____

Use with Unit 3.   Science Content Support   CS 69

Name _____
Date _____

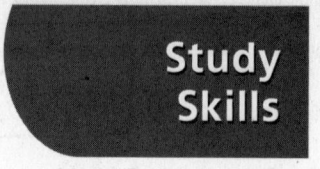

Unit 3, Lesson 2

Lesson 2—What Affects Survival?

Anticipation Guide

An anticipation guide can help you anticipate, or predict, what you will learn as you read.

- Look at the section titles for clues.
- Preview the Reading Focus Skill question at the end of each section. Use what you know about the subject of each section to predict the answers.
- Read to find out whether your predictions were correct.

As you read each section, complete the anticipation guide below. Predict answers to each question and check to see if your predictions were correct.

What Affects Survival?		
Adaptation and Accommodation		
Reading Focus Skill	Prediction	Correct?
What kinds of changes can lead to the development of adaptations?		
Tropical Rain Forests		
Reading Focus Skill	Prediction	Correct?
How does camouflage help animals?		
Coral Reefs		
Reading Focus Skill	Prediction	Correct?
What do you think happens when the algae living in coral die?		
Deserts		
Reading Focus Skill	Prediction	Correct?
How do adaptations help desert animals survive?		

CS 70 Science Content Support

Use with Unit 3.

Name _____
Date _____

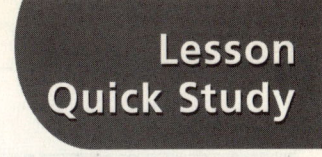

Unit 3, Lesson 2

Lesson 2—What Affects Survival?

1. **Investigation Skill Practice–Draw Conclusions**

    The winds from a hurricane blew trees down, and the stormy ocean washed away many plants from the beach. After the storm passed, Janet noticed that there were not as many animals near the ocean's shore. After some of the plants had grown again, Janet noticed that the animals had returned. Draw conclusions about why the animals might have been missing for two years.

    _____
    _____
    _____
    _____

2. **Reading Focus Skill Practice–Cause and Effect**

    Read the selection. Describe the cause and effect of the of the white moths' adaptation to the factories.

    When factories were first built in England, they were very dirty and smoky. Black soot covered trees and buildings. Scientists noticed how a little white moth adapted to the black trees. Scientists knew that the white moths were mostly born white, but sometimes one was black or grey. When the trees became black, the white moths were eaten by birds before they could lay their eggs. The black moths were able to hide better, and the birds could not see them. They grew old enough to lay their eggs. Soon the white moths had adapted and survived as black moths.

    _____
    _____
    _____
    _____

Use with Unit 3.   (page 1 of 2)   Science Content Support   CS 71

Name _____

Science Concepts

3. Using the chart below, give one example of an adaptation that helps an organism survive in each ecosystem.

Different Ecosystems		
Tropical Rain Forest	Coral Reef	Desert
Adaptation Example:	Adaptation Example:	Adaptation Example:

4. Tell what ecosystem each plant or animal belongs to. Then tell how it has adapted to its environment.

clown fish _____

gila monster _____

orchids _____

toucan _____

kangaroo rat _____

5. Give an example of an animal that uses camouflage to survive.

_____

_____

6. Give an example of an animal that has made an accommodation in order to survive in its environment.

_____

_____

CS 72 Science Content Support   (page 2 of 2)   Use with Unit 3.

Name _____

Date _____

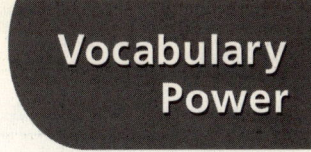

Unit 3, Lesson 3

Lesson 3—What Is Interdependence?

A. Context Clues

Read the sentences. Write the meaning of the underlined word on the line.

1. The scientists noticed the <u>interdependence</u> of fish because they depended on one another to survive.

   _____

2. When the bees <u>pollinate</u> the huge orange flowers of the pumpkin vine, they take the pollen from the male to the female part of the plant.

   _____

3. Cleaner shrimp have a <u>relationship</u> with fish because they get rid of the fish's parasites by eating the parasites.

   _____

B. Writing Sentences

On the lines below, write three sentences that use the words in the box correctly.

| interdependence | pollinate | relationship |

_____

_____

_____

Use with Unit 3.   Science Content Support   CS 73

Name _____
Date _____

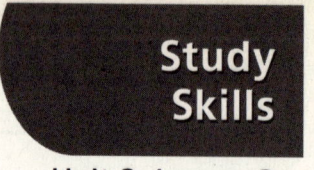

Unit 3, Lesson 3

Lesson 3—What Is Interdependence?

Write to Learn

Writing about what you read can help you better understand and remember information.

- Writing down the information that you learn from each lesson leads you to think about the information.
- Writing your own response to the new information makes it more meaningful to you.

As you read the lesson, pay attention to new and important information. Keep track of the information and your responses in the log below.

What Is Interdependence?	
What I Learned	**Personal Response**
Some animals rely on one another to live.	Knowing that animals protect one another reminds me to respect the habitat of other living creatures.

Use with Unit 3.

Name _____

Date _____

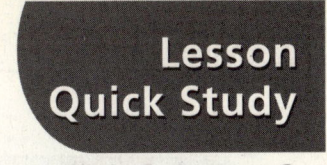

Unit 3, Lesson 3

Lesson 3—What Is Interdependence?

1. **Investigation Skill Practice–Predict**

   Julio lived in the city. But he wanted a garden. He found a small plot of dirt. He got permission from the owner to plant tomatoes, jalapeño peppers, basil, and beans. He also planted lots of flowers along the outside of the garden. He was disappointed, because his plants did not grow any fruit. The next year he had an idea. He made a wooden hive for bees. Why do you think his garden did not grow fruit the first year? What do you predict will happen to his garden the next year?

   _____

   _____

2. **Reading Skill Practice–Compare and Contrast**

   Read the selection. Compare and contrast the different types of zoos.

   While all zookeepers think about the best interest of their animals, some zoos plan their exhibits differently. At some zoos, the planners make individual habitats for each species. For example, the giraffes may have a fenced area with tall trees and tall fences. The zebras may have an open plain where they can run. Other zookeepers make different plans. They know that many species rely on one another to survive in the wild. These zookeepers put interdependent species together. They place the zebras, the giraffes, and the gazelles together. The animals do not compete for the same food, so they get along well in the same zoo habitat.

   _____

   _____

Name _____

3. **Science Concepts**

Read the following story and answer the questions that follow about the interdependence of the animals and plants in Renee's backyard.

A family of squirrels and several groups of birds lived in Renee's backyard. There were a flock of crows, two noisy blue jays, and several snakes. Renee's cat also lived in her backyard. Renee had planted a cherry tree and a holly bush that grew berries in the winter. There was a tall oak that dropped acorns in the fall and a big pine tree that dropped pine cones. Bees built a hive in the hole of the oak tree. There were also several big leafy bushes along the edge of her backyard. A hole in the ground was a home to hundreds of ants.

A. How do the animals depend on the plants?

_____
_____
_____
_____

B. How do the plants depend on the animals?

_____
_____
_____
_____

C. How do the animals depend on one another?

_____
_____
_____
_____

Name _____
Date _____

Unit 3, Lesson 4

Lesson 4—What Are Microorganisms?

A. Etymology

Prefixes are word parts that are added to root words to change their meaning.

Prefix	Origin	Meaning	Words
Micro	Greek	small	microscope, microorganism
Proto	Greek	first	protist, protozoans
Phyto	Greek	coming from a plant	phytoplankton

Using the prefixes in the chart, match the words below to their definitions.

1. _____ protist
2. _____ microscope
3. _____ phytoplankton
4. _____ microorganism

a. a science tool that makes very tiny things look bigger

b. a one-celled organism that may share traits with plants or animals

c. plankton that is made up of plants, such as algae

d. organisms that are too small to see without a microscope

B. Write Sentences

Write two sentences that use the following vocabulary words.

| bacteria | mold |

_____

_____

Use with Unit 3.   Science Content Support   CS 77

Name _____
Date _____

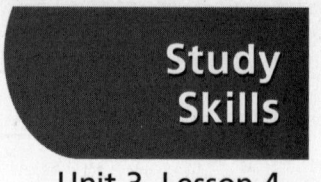

Unit 3, Lesson 4

Lesson 4—What Are Microorganisms?

Use a K-W-L Chart

A K-W-L chart can help you focus on what you already know about a topic and what you want to learn about it.

- Use the K column to list what you already know about microorganisms.
- Use the W column to list what you want to learn about microorganisms.
- Use the L column to list what you have learned about microorganims from your reading.

Complete the K-W-L chart as you read this lesson.

Microorganisms		
What I <u>K</u>now	What I <u>W</u>ant to Learn	What I <u>L</u>earned
• I know some microorganisms are harmful to humans.	• Are all microorganisms harmful?	• _____ • _____ • _____ • _____
• _____ • _____ • _____ • _____ • _____ • _____ • _____ • _____	• _____ • _____ • _____ • _____ • _____ • _____ • _____ • _____	• _____ • _____ • _____ • _____ • _____ • _____ • _____ • _____

Science Content Support

Use with Unit 3.

Name _____

Date _____

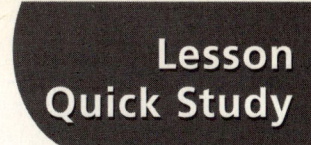

Lesson Quick Study

Unit 3, Lesson 4

Lesson 4—What Are Microorganisms?

1. **Investigation Skill Practice—Infer**

   Yeast makes bread rise because it feeds on sugar in the dough and produces bubbles of carbon dioxide. The carbon dioxide gets trapped in the layers of bread dough and then baked in the bread. From what you know about bread, make inferences to explain what would happen if there were no sugar in the dough.

   _____
   _____
   _____
   _____

2. **Reading Skill Practice—Cause and Effect**

   Read the selection. Describe the cause and effect of an oil leak.

   One winter day, an enormous oil tanker hit a sandbar. The oil tanks sprang a leak. Soon the ocean waves were covered in oil. The oil was very bad for the animals of the sea. Birds became covered in oil and could not fly. Whales could not breathe through oily water. When oil got on a seal's fur, it quickly collected dirt and the seal could not stay warm. The oil company discovered how to clean up the oil spill. They used a microorganism that eats oil. This microorganism lives on the oil until it is gone. Then it dies and decays in the water like any other organism.

   _____
   _____
   _____
   _____

Name _____

3. Science Concepts

Use what you learned in the lesson to fill in the chart below.

	Harmful Microorganisms	Helpful Microorganisms
Where do they live?		
What do they do?		
What are some examples?		

Science Content Support

Name _____

Date _____

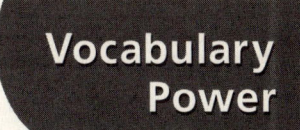

Unit 4, Lesson 1

Lesson 1—How Are Minerals Identified?

A. Graphic Organizer

Fill in the chart with information about each word on the left. The first one is done for you.

	Ways to Identify Minerals	
	What is it?	**Give an example.**
color	one property of minerals, but not a good way to identify minerals	Pyrite, or fool's gold, has a gold color.
streak		
luster		
hardness		

B. Writing Sentences

Write three sentences using the words in the box.

| mineral | luster | streak |

_____
_____
_____
_____
_____

Use with Unit 4.   Science Content Support   **CS 81**

Name _____
Date _____

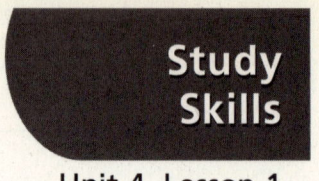

Unit 4, Lesson 1

Lesson 1—How Are Minerals Identified?

Take Notes

Taking notes can help you remember important ideas.

- Write down important facts and ideas. Use your own words. You do not have to write in complete sentences.
- One way to organize notes is in a chart. Write down the main ideas in one column and facts and details in another.

As you read this lesson, use the chart below to take notes.

How Are Minerals Identified?	
Main Ideas	**Facts**
• Minerals are made from materials that were never alive. • _____ _____ • _____ _____ • _____ _____ • _____ _____	• Minerals can form deep in the Earth, in caves, from seawater, from underground water, and in geodes. • _____ _____ • _____ _____ • _____ _____ • _____ _____

Science Content Support

Use with Unit 4.

Name _____

Date _____

Lesson Quick Study

Unit 4, Lesson 1

Lesson 1—How Are Minerals Identified?

1. **Investigation Skill Practice–Classify**

   Look at the hardness chart. Complete the chart by writing the minerals underneath the hardness description. Choose from the minerals listed below.

1 Talc	4 Fluorite	7 Quartz	10 Diamond
2 Gypsum	5 Apatite	8 Topaz	
3 Calcite	6 Orthoclase	9 Ruby	

   Hardness Chart

Minerals that can be scratched by a fingernail: hardness < 2.5	Minerals that can be scratched by a copper penny: hardness ≤ 3	Minerals that can be scratched by a steel nail: hardness ≤ 5.5	Minerals that can be scratched by glass: hardness ≤ 6	Minerals that cannot be scratched by any of these

2. **Reading Focus Skill Practice–Main Idea and Details**

   Read the selection. Underline the main idea. Circle 3 details.

   Calcite deposits have created huge colorful formations at Luray Caverns in Virginia. One formation looks like double pipes with yellow stripes. They rise from the floor of the cave. In another area, long, red, bumpy calcite icicles hang from the cave ceiling. In one place, the calcite is white and looks like a white veil.

Name _____

Science Concepts

3. Read the chart and answer the questions about identifying minerals.

Mineral	Hardness	Luster	Streak	Color	Cleavage/Fracture
Calcite	3	Glassy	colorless or white	colorless, white, or various colors	easy cleavage and curved fracture
Copper	2.5–3	Metallic	copper orange	copper orange to dark red	jagged fracture

A. Can you scratch calcite with copper?
_____

B. Which mineral is harder, calcite or copper?
_____
_____

C. Are the minerals similar or different in the way they break? Explain.
_____
_____

D. What happens when you break copper?
_____

E. What color does calcite make when you scrape it on a tile?
_____

F. Is the streak of copper the same as its color?
_____

Name _____

Date _____

Extra Support

Unit 4, Lesson 1

Lesson 1—How Are Minerals Identified?

Use Properties to Identify Minerals

Minerals are nonliving solid substances. They occur naturally and have a repeating structure. Minerals can be identified by their properties, such as color, streak, hardness, luster, cleavage, and fracture.

This table summarizes the properties of many common minerals.

Mineral Identification Table

Mineral	Hardness	Luster	Streak	Color	Other
Talc	1	nonmetallic	white	white, greenish to gray	fractures into clumps
Graphite	1–2	metallic	black	gray to black	slippery, used in pencil leads
Mica	2–2.5	nonmetallic	none	dark brown, black or silver-white	cleaves into thin sheets
Galena	2.5–3	metallic	gray	lead-gray	lead ore
Gold	2.5–3	metallic	golden yellow	yellow	
Calcite	3	nonmetallic	white	colorless, white	
Hornblende	5–6	nonmetallic	none	dark green to black	black flecks in granite
Hematite	5–6.5	metallic or nonmetallic	reddish brown	silver-gray or red	iron ore
Feldspar	6	nonmetallic	none	colorless, beige, pink	major part of pink granite
Magnetite	6	metallic	black	black	
Pyrite	6–6.5	metallic	greenish black	brassy yellow-gold	
Quartz	7	nonmetallic	none	colorless, white, rose, smoky purple, brown	

Use with Unit 4.   (page 1 of 4)   Science Content Support   CS 85

Name _____

Mohs Hardness Scale

Mineral	Hardness
Talc	1
Gypsum	2
	2.5
Calcite	3
	3.2
Fluorite	4
Apatite	5
	5.5
Feldspar	6
	6.5
Quartz	7
Topaz	8
Corundum	9
Diamond	10

1. Besides color, identify three properties of magnetite.

_____

2. Suppose you find a dark green mineral that leaves no mark on a streak plate. It scratches fluorite, but not quartz. What is the mineral?

_____

3. What is the hardness of a metallic gray mineral with a gray streak? Explain.

_____

4. You test a nonmetallic, light pink mineral for hardness and find that it falls in the 5.5–6 range. Do you have enough information to identify it? If so, name it and explain your answer. If not, explain why not.

_____

_____

Name _____

5. You find a sample of a metallic brassy yellow mineral. Describe two ways you can tell if it is really gold.

_____
_____
_____
_____

6. Quartz has no streak. Look at its other properties and suggest a reason why that may be so.

_____
_____

7. True or False: Metallic minerals tend to be harder than nonmetallic ones. Use the information in the chart to justify your answer.

_____
_____
_____
_____
_____

Name _____

8. Which mineral are you probably using to answer these questions? Why?
_____
_____

9. According to the table, which mineral occurs in the most colors?
_____

10. Which minerals in the table can be scratched by a penny? How do you know?
_____
_____

11. Which minerals in the table can be scratched by a knife blade but not glass? How do you know?
_____
_____
_____

12. Which minerals in the table can quartz scratch? Why?
_____

13. You have two mineral samples from the table. They both are white with a white streak. What property can you use to distinguish them? What are they?
_____
_____

14. Name two minerals from the table that are used to make things. Give an example of how each is used.
_____
_____

Name _____

Date _____

Vocabulary Power

Unit 4, Lesson 2

Lesson 2—How Are Rocks Identified?

A. Explore Word Meanings

Think about the meaning of the underlined words. Then write an answer to each question.

1. An <u>igneous rock</u> forms when melted rock cools and hardens. Where are you more likely to find an igneous rock—near a volcano or near an iceberg? Why?

   _____

2. A <u>sedimentary rock</u> forms when pieces of rocks are broken down and moved. The broken down rocks pile up in layers and are squeezed together. Which would cause a rock to break down faster—wind or sunlight? Why?

   _____

3. A <u>metamorphic rock</u> forms when high temperature and pressure change an existing rock. What is more likely to cause a metamorphic rock to form—ocean waves or mountain building? Why?

   _____

4. A <u>rock</u> is a naturally occurring solid made up of minerals. Which of the following is a rock—oil or granite? Why?

   _____

5. The <u>texture</u> is a property of a rock that you can use to classify the rocks you find. Which of these descriptions tell the texture of a rock—small-grained or red?

   _____

Use with Unit 4.   Science Content Support   CS 89

Name _____
Date _____

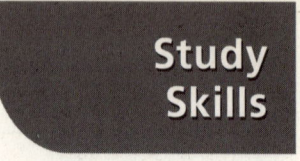

Unit 4, Lesson 2

Lesson 2—How Are Rocks Identified?

Understanding Vocabulary

Using a dictionary can help you learn new words that you find as you read.

- A dictionary shows all the meanings of a word and tells where the word came from.
- You can use a chart to list and organize unfamiliar words that you look up in a dictionary.

As you read this lesson, look up unfamiliar words in the dictionary. Add them to the chart below. Fill in each column to help you remember the word's meaning.

Word	Syllables	Origin	Definition
rock	RAHK	Middle English, Old French, and medieval Latin	mineral matter formed by the action of heat, water, etc.
igneous			
metamorphic			
sedimentary			

CS 90 Science Content Support

Use with Unit 4.

Name _____
Date _____

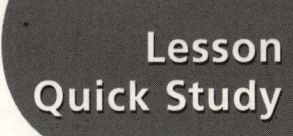

Unit 4, Lesson 2

Lesson 2—How Are Rocks Identified?

1. **Investigation Skill Practice – Infer**

   When a forest fire burned near a town, it destroyed several houses. After the fire went out, a rescue worker looked through the remains of a house. She found metal pieces in the coals that had been toys and kitchen equipment. She noticed that the metal had become mangled and twisted. What inferences can you make to explain how the metal pieces changed?

   _____
   _____

2. **Reading Skill Practice – Compare and Contrast**

   Read the selection. Compare and contrast Raffi's two rocks.

   When Raffi and his brothers went on vacation, they hiked on a mountain. Raffi collected several rocks to study at home. He compared two rocks. Both were black. But he noticed that one rock was lightweight and was filled with many holes. Another rock was much heavier even though it was about the same size. He noticed that layers of the second rock could be split off in sheets. He determined that the first rock was igneous rock. The second was sedimentary.

   _____
   _____
   _____

Name _____

Science Concepts

3. Read the list of descriptions. Place each description under the type of rock that it describes.

Some of these rocks form underground and tend to have large crystals.

These are formed from small pieces of rock that have been "glued" together.

These are hard and do not usually have layers.

These rocks are often banded because of the high temperature and pressure that form them.

These rocks are changed by high temperature and pressure.

These rocks can form near volcanoes, where there is a lot of heat.

This type of rock usually shows layers.

Igneous Rocks

_____

_____

Sedimentary Rocks

_____

_____

_____

Metamorphic Rock

_____

_____

_____

_____

_____

_____

Name _____
Date _____

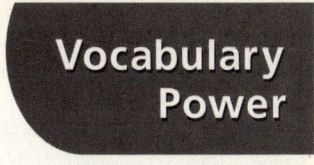

Unit 4, Lesson 3

Lesson 3—What Is the Rock Cycle?

A. Context Clues

Study the words and their definitions below. Then, using the context clues, fill in the blanks with one of the words. Use all of the words once.

deposition the process of depositing or setting down sediment after it is eroded

erosion the process of carrying sediment away from its source

lava molten rock that is on Earth's surface

magma molten rock that is beneath Earth's surface

rock cycle an endless process in which rocks are changed from one type to another by weathering, erosion, and high temperature and pressure

weathering the process of wearing away rocks

Fill in the blanks with one of the words above. Use all of the words once.

1. The molten rock that remains under the Earth's surface is _____.

2. Molten rock that comes to the surface of the Earth is called _____.

3. The slow, continuous process of changing rocks from one type to another is called the _____.

4. Water, wind, and rain, all wear away rocks during the process called _____.

5. While water washes down a slope, it carries dirt and soil away with it. This process is called _____.

6. At the bottom of a slope, you can often see where sediments have been set down through the process of _____.

Use with Unit 4.   Science Content Support   CS 93

Name _____

Date _____

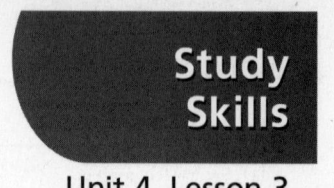

Unit 4, Lesson 3

Lesson 3—What Is the Rock Cycle?

Skim and Scan

Skimming and scanning are two ways to learn from what you read.

- To skim, quickly read the lesson title and the section titles. Look at the visuals, or images, and read the captions. Use this information to identify the main topics.
- To scan, look quickly through the text for specific details, such as key words or facts.

Before you read this lesson, skim the text to find the main ideas. Then look for key words. If you have questions about a topic, scan the text to find the answers. Fill in the chart below as you skim and scan.

What Is the Rock Cycle?	
Skim	**Scan**
Lesson Title:	Key Words and Facts:
Main Idea:	
Section Titles:	
Visuals:	

CS 94 Science Content Support

Use with Unit 4.

Name _____
Date _____

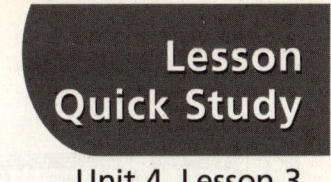

Lesson
Quick Study

Unit 4, Lesson 3

Lesson 3—What Is the Rock Cycle?

1. **Investigation Skill Practice–Model**

   Kaylie used a hose and a pile of dirt to make a model of the process of erosion. She let water flow from the hose over the dirt. The water made a deep gully in the dirt pile. She could tell that the water did not destroy the dirt. The water just moved the dirt from one place to another. She recorded her observations in her notebook.

   What equipment did Kaylie use for her model?
   _____

   What does Kaylie's model show?
   _____

   What can you learn from a model that you cannot learn from a book?
   _____

2. **Reading Skill Practice–Cause and Effect**

   Read the selection. Describe the cause and effect of the cracking rock.

   Many factors cause weathering. It happens all around us. Suppose a large rock has a small crack. Now suppose a bird carries a seed as it flies above the rock. The seed drops from the bird's beak and falls into the rock crack. With time, the seed grows into a plant. The growing plant's roots expand, breaking the rock even further. Time continues to pass and winter arrives. It rains. Water gets into the crack. The water freezes. As the water freezes, it expands. The crack continues to expand. Before you know it, the rock with a small crack has a much larger crack.

   _____
   _____
   _____

Use with Unit 4.

Name _____

Science Concepts

3. Read the steps in the box below. Put the steps in the order they might occur in the rock cycle.

> High temperature and pressure act on the sandstone, turning it into a metamorphic rock.
>
> The new rock hardens into an igneous rock.
>
> Sandstone gets pushed deep into Earth's crust.
>
> After time, the new rock cools.
>
> The metamorphic rock melts after undergoing even more heat and pressure.

Step 1
_____

Step 2
_____

Step 3
_____

Step 4
_____

Step 5
_____

Name _____
Date _____

Vocabulary Power

Unit 5, Lesson 1

Lesson 1—What Causes Changes to Earth's Surface?

A. Explore Word Meaning

Match the words from the list on the right with their correct definitions.

creep
dune
earthquake
landslide
lava
volcano

1. _____ the shaking of Earth's surface caused by movement of rock in the crust

2. _____ the sudden movement of rock and soil downhill

3. _____ the slow movement of soil or rock downhill

4. _____ a mountain that forms as molten rock flows through a crack onto Earth's surface

5. _____ molten rock on Earth's surface that flows from a volcano

6. _____ a large mound of sand piled by wind

B. Analogies

Choose one of the words from the box to complete the analogies.

lava
dune
glacier
creep

7. _____ is to *volcano* as *toothpaste* is to *toothpaste tube*.

8. _____ is to *sand* as *forest* is to *trees*.

9. _____ is to *ice* as *lake* is to *water*.

10. _____ is to *slow* as *race car* is to *fast*.

Use with Unit 5.

Science Content Support   CS 97

Name _____
Date _____

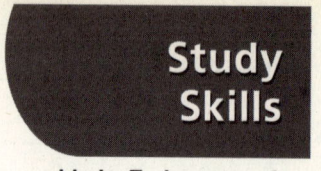

Study Skills

Unit 5, Lesson 1

Lesson 1—What Causes Changes to Earth's Surface?

Use a K-W-L Chart

A K-W-L chart can help you focus on what you already know about a topic and what you want to learn about it.

- Use the K column to list what you already know about volcanoes, earthquakes, and landslides.
- Use the W column to list what you want to learn about the topic.
- Use the L column to list what you have learned about the topic from your reading.

Complete the K-W-L chart as your read this lesson.

Volcanoes, Earthquakes, and Landslides		
What I Know	**What I Want to Learn**	**What I Learned**
• I know that volcanoes, earthquakes, and landslides can be very dangerous.	• What causes these changes to the Earth?	• _____ • _____ • _____ • _____
• _____ • _____ • _____ • _____ • _____ • _____ • _____ • _____ • _____	• _____ • _____ • _____ • _____ • _____ • _____ • _____ • _____ • _____	• _____ • _____ • _____ • _____ • _____ • _____ • _____ • _____ • _____

Science Content Support

Use with Unit 5.

Name _____

Date _____

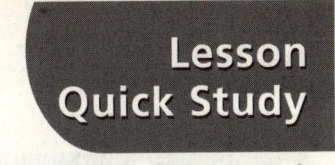

Lesson Quick Study

Unit 5, Lesson 1

Lesson 1—What Causes Changes to Earth's Surface?

1. **Investigation Skill Practice–Infer**

   During the Ice Age, huge glaciers scraped the Earth, including California. They left ridges of soil and rock when they melted. Make inferences to tell how the climate in California has changed. What was it like during the Ice Age? Explain why there are no glaciers today.

   _____
   _____
   _____
   _____

2. **Reading Focus Skill Practice–Compare and Contrast**

   Read the selection. Compare and contrast the different kinds of earthquakes.

   All earthquakes are caused by a shift in the Earth's crust. However, each earthquake changes the Earth in different ways. For example, some earthquakes are very strong. They leave huge cracks in the land. They shake houses and knock over buildings. Other earthquakes are much weaker. They make a rumbling noise for miles around. They might shake the Earth a little bit. When the rumbling is over, there are no big cracks. These earthquakes are smaller and less destructive.

   _____
   _____
   _____
   _____

Name _____

Science Concepts

3. For each type of natural event, answer the questions in the chart.

	What is it?	Does it cause fast or slow change to the Earth's surface?	What change does it make to the Earth's surface?
earthquake			
landslide			
glacier			
volcano			
creep			
desert pavement			

4. Answer the questions.

Is a landslide more likely to happen in a desert or in an area with plenty of rain?

_____

How can ash from a volcano be both good and bad?

_____

Name _____
Date _____

Unit 5, Lesson 2

Lesson 2—What Causes Weathering?

A. Word Families

Look at each word root. Write a word from the box that belongs to the same word family.

| weathering | chemical | exfoliation | soil |
| mechanical | abrasion | oxidation | |

1. weather _____

2. machine _____

3. chemistry _____

4. abrade _____

5. foliate _____

6. oxygen _____

B. Context Clues

Complete each sentence or sentences with the correct word from the box.

7. The visitors could see the damage that acid rain had done. The rocks had been damaged by _____ weathering.

8. To make your _____ rich and fertile, you must add dead organisms and water to the weathered rock.

9. The glaciers of the Ice Age caused the _____ of the scraped boulders.

10. As we hiked, we could tell that the sharp rocks were freshly broken. They had not been through the _____ of hundreds of years of wind, rain, and ice.

Use with Unit 5.　　　　　　　　　　　　　Science Content Support　　CS 101

Name _____
Date _____

Unit 5, Lesson 2

Lesson 2—What Causes Weathering?

Anticipation Guide

An anticipation guide can help you anticipate, or predict, what you will learn as you read.

- Look at the section titles for clues.
- Preview the Reading Focus Skill question at the end of each section. Use what you know about the subject of each section to predict the answers.
- Read to find out whether your predictions were correct.

As you read each section, complete the anticipation guide below. Predict answers to each question and check to see if your predictions were correct.

What Causes Weathering?		
Mechanical Weathering		
Reading Focus Skill	Prediction	Correct?
How does abrasion cause rocks to break down?	_____ _____ _____ _____	✓
Chemical Weathering		
Reading Focus Skill	Prediction	Correct?
How does chemical weathering cause change in rock?	_____ _____ _____ _____	✓

CS 102 Science Content Support

Use with Unit 5.

Name _____
Date _____

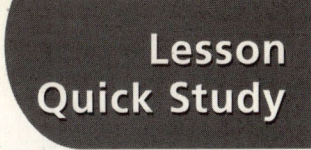

Unit 5, Lesson 2

Lesson 2—What Causes Weathering?

1. Investigation Skill Practice–Predict

A sculptor carved two statues for the city where he was born. Both were the same size. However, one was kept at a museum under dim lights. It was warm in the winter and cool in the summer. The other statue was placed in the city park. In the summer, birds nested on it. In the winter, icicles formed on its corners. What do you predict will happen to each statue?

_____
_____
_____
_____

2. Reading Focus Skill Practice–Cause and Effect

Read the selection. Describe the cause and effect of the doctor's visitors.

Sherlock Holmes lived in an apartment on Baker Street. Before Holmes moved in, the apartment had been a doctor's office. Holmes noticed that the front step of his new apartment was deeply grooved from the scraping of many feet that had climbed the stair. Holmes deduced that the doctor must have been a very good one, since so many people had come to him when they were sick.

_____
_____
_____

Use with Unit 5.   (page 1 of 2)   Science Content Support   CS 103

Name _____

Science Concepts

3. **Each statement below has a mistake. Rewrite the sentence correctly.**

   Mechanical weathering is when a machine, such as a car, drives over the rock.

   _____

   _____

   Ice is not as strong as rock, so it cannot change rocks at all.

   _____

   Water and wind change rocks by breaking them in two.

   _____

   Rocks get smaller when they are hot. They get bigger when they are cold.

   _____

   Rocks always weather smoothly and evenly from every surface.

   _____

   When iron mixes with oxygen, it makes the rock stronger.

   _____

   All rocks weather at the same rate.

   _____

   Chemical weathering happens fastest in cold, dry deserts.

   _____

   Chemical weathering only happens on the surface of the ground, not underground.

   _____

   _____

Name _____
Date _____

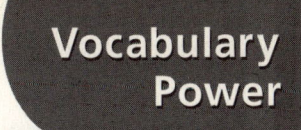

Unit 5, Lesson 3

Lesson 3—How Does Moving Water Shape the Land?

Context Clues

Complete the story by filling in each blank. Use a word from the box.

| beach | deposition | sea arch |
| runoff | erosion | transport |

When Benjamin and his sister built a sand castle on the _____, they learned a lot about how water shapes the land. They piled sand into a huge mountain for the main part of the castle. Then they made a moat around the castle. They poured water into the moat and made a gully for a river. It would flow from the castle to the ocean. They expected the water to flow over the sand like _____ over dirt. But instead, the water usually sunk in to the soft sand. Sometimes it washed away part of their wall. Soon there was a wide _____ of sand on the lower beach where their river had dropped the sand. When the tide came in, the ocean waves began to beat against their castle. At first, the waves caused _____ of the moat. Then it cut into the base of their castle, washing it away. It created a _____. Then big chunks of the castle mountain fell with a splash into the moat. The waves kept washing up to _____ their sand out into the ocean. Benjamin and his sister decided that the next castle would be built much farther from the waves.

Use with Unit 5.   Science Content Support   CS 105

Name _____
Date _____

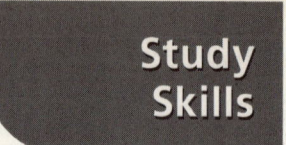

Unit 5, Lesson 3

Lesson 3—How Does Moving Water Shape the Land?

Write to Learn

Writing about what you read can help you better understand and remember information.

- Writing down the information that you learn from each lesson leads you to think about the information.
- Writing your own response to the new information makes it more meaningful to you.

As you read the lesson, pay attention to new and important information. Keep track of the information and your responses in the log below.

How Does Moving Water Shape the Land?	
Rivers Shape the Land	
What I Learned	Personal Response
Ocean Waves Shape the Land	

Science Content Support — Use with Unit 5.

Name _____

Date _____

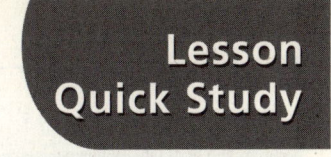

Unit 5, Lesson 3

Lesson 3—How Does Moving Water Shape the Land?

1. **Investigation Skill Practice–Make a Model**

Ingrid wanted to make a model to show how ocean waves change the land. She plans to observe what happens when she throws large cups of water at a pile of pebbles, rocks, and sand. Answer the questions about how Ingrid should set up her model.

What materials does Ingrid need?
_____
_____

What will Ingrid show with her model?
_____
_____

2. **Reading Skill Practice–Main Idea and Details**

Read the selection. Underline the main idea. Write 3 details on the lines below.

 The Grand Canyon is amazing for three reasons. First, the canyon is very deep. The Colorado River has carved a deep gully more than one mile below the canyon rim. Second, the river is powerful. It has some of the most powerful rapids in the world. Third, the rock that has been carved by the river is beautiful. It shows stripes of red, yellow, and brown.

_____
_____

Use with Unit 5.   (page 1 of 2)   Science Content Support   CS 107

Name _____

Science Concepts

3. **Read the statements and answer the questions.**

   Statement 1: In some areas, huge waves crash against rocky cliffs.

   How are the rocks being changed by the waves?

   _____

   _____

   What will happen to the rocks?

   _____

   _____

   Statement 2: At the place where a river empties into the ocean, the river widens, and there is a huge delta.

   What speed is the river going when it meets the ocean?

   _____

   _____

   How did the delta form?

   _____

   _____

   Statement 3: Some distance from the shore, there is a strip of sand called a barrier island.

   How did it form?

   _____

   _____

   Will it stay there forever, or will it change?

   _____

   _____

VOCABULARY GAMES and CARDS

Contents

Vocabulary Games

Guess the Word CS110

Word Ladder CS110

Hidden Words CS111

Crossed Words CS111

Name that Word CS112

Red Light, Green Light... CS112

Vocabulary Cards CS113–CS175

Vocabulary Games

You can use the vocabulary cards on pages CS113–CS175 to play these games. The cards are provided for each chapter in your science textbook. Each card has a word on one side and the word's definition on the back. For some of these games, you may need to keep the definition hidden from view.

Guess the Word

You will need
vocabulary cards, paper and pencil

Grouping large groups or pairs

1. Form two teams. Each player must have a partner. One player in each pair is the clue giver, and one is the guesser.

2. Teams take turns playing. The first clue giver draws a word card and gives the guesser one clue at a time. Count clues to keep score.

3. After the word is guessed, play is passed to a pair on the other team. Use all the cards. The lowest score wins.

Word Ladder

You will need
vocabulary cards, tabletop

Grouping groups or partners

1. Place the cards in a pile, hiding the definitions.

2. Player 1 chooses a card, reads the word, and says the word's meaning. That player then turns the card over to check his or her answer.

3. If the meaning is correct, the word is placed near the edge of a tabletop. Player 1 continues until a word is missed. If more words are guessed correctly, the words are added to Player 1's ladder. If a word is missed, the card is returned to the pile. It is then the next player's turn.

4. The player who has formed the tallest ladder is the winner!

Hidden Words

You will need

vocabulary cards, paper and pencil

Grouping Whole class or large groups; small groups; pairs

1. Choose a word from the vocabulary cards to hide in a sentence. For example, the word *heat* is hidden in the following sentence: Mitc<u>h</u> <u>eat</u>s ice cream.

2. Write a sentence with a hidden word. Exchange papers with a classmate to find each other's hidden words.

Crossed Words

You will need

vocabulary cards, grid paper, and pencil

Grouping small groups or pairs

1. Place the cards face-up on a table so that each word can be seen. Choose one word for your crossword puzzle. Write that word vertically on the grid. Identify it by writing a number 1 in the box with the first letter, just as on a crossword puzzle. Use the word's definition on the back of the card to help write a clue for this starter word.

2. Choose a second word to add to the grid. Be sure it shares a letter with the first word written. Attach it to the first word by writing the second word horizontally on the grid. Identify it by writing a number 2 in the box with the first letter of that word.

3. Continue to attach words to the puzzle and number each word. Write clues for each of the numbered words.

4. Give a partner a blank grid with spaces numbered to match your puzzle. To help your partner solve the puzzle, shade each square of the grid that doesn't contain a letter. Challenge your partner to solve the puzzle by reading the clues and guessing each word.

Vocabulary Games

Name That Word

You will need
2 identical sets of word cards; paper and pencil

Grouping small groups of at least five

1. One player is named as the "host." Players pair off into two teams. Each team has a set of word cards in the same order, placed facedown.

2. The host asks one person from each team to draw a card, checking to make sure both are looking at the same word. The player from Team A who saw the word goes first. He or she gives a one-word clue about the word. Team members then try to guess the word.

3. If the word is guessed, the host gives Team A a point. If the word is not guessed, Team B gets a turn. The player from Team B who saw the word gives a second one-word clue about the word. Team B then tries to guess the word. If the word is not guessed after five rounds, no team scores a point, and the host reveals the word to both teams.

4. After all of the cards have been drawn, the team with the most points wins.

Red Light, Green Light

You will need
vocabulary cards sorted by chapter, science textbook

Grouping large or small groups

1. Each player has a set of the same word cards. One person is the host of the game and does not play.

2. The host holds up one word for everyone to see. Each player then pulls that word card out of his or her pile. All players open their books to the chapter from which the word has been taken.

3. The host calls out "green light." Each player quickly looks for the word in a sentence. When a player finds the word, the word card marks the page and the book is closed. That player calls out "red light." Play then stops.

4. The player who stopped the game reads the sentence that holds the word. He or she scores one point. Play continues with the next word.

accommodation

adaptation

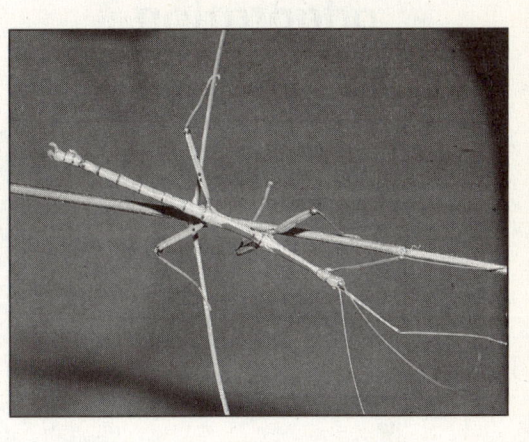

attract

axis

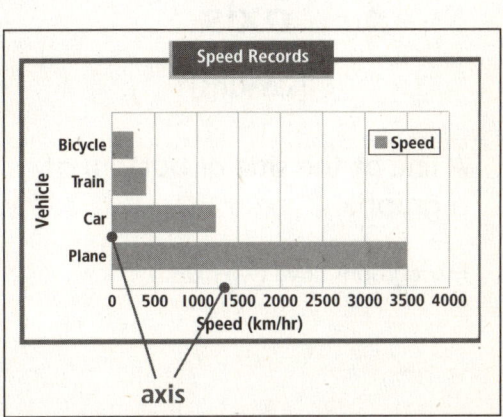

Vocabulary Cards

Science Content Support CS 113

accommodation
[uh•kahm•uh•DAY•shuhn]

A learned behavior that helps the individual members of a species survive.

Wearing gloves is an *accommodation* that helps people survive in cold climates.

adaptation
[a•duhp•TAY•shuhn]

A body part or behavior that helps an organism survive.

This insect's stick-like body is an *adaptation* that makes it look like part of a tree.

attract
[uh•TRAKT]

To pull toward.

Magnets *attract* iron.

axis
[AKS•uhs]

A line at the side or bottom of a graph.

Most graphs have two *axes*.

bacteria

battery

biome

boulder

Vocabulary Cards

Science Content Support CS 115

battery
[BAT•uh•ree]

An energy storage device that uses chemical energy to produce a flow of electric current.

We use *batteries* to power some toys.

bacteria
[bak•TEER•ee•uh]

A certain type of microorganism.

Some *bacteria* can make you sick.

boulder
[BOHL•der]

A large, rounded rock larger than 256 millimeters (10 in.) in diameter.

Boulders are usually heavy and hard to move.

biome
[BY•ohm]

A large area that has a similar climate and similar ecosystems in all parts of it.

A deciduous forest is one kind of *biome*.

bulb
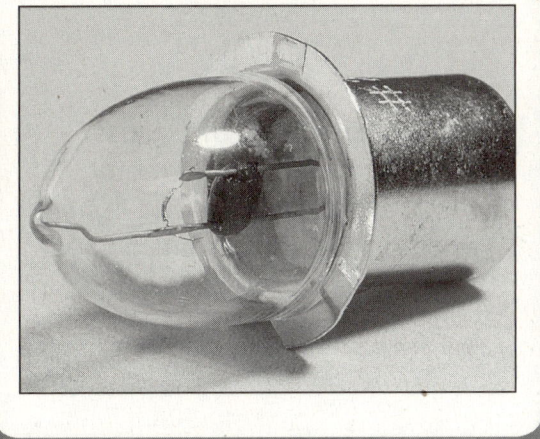

carnivore

chemical change

chemical weathering

carnivore

[KAR•nih•vawr]

An animal that eats only other animals.

Carnivores have sharp teeth to help them eat meat.

bulb

[BUHLB]

A device made up of a glass globe, a filament, and metal, which produces heat and light when an electric current passes through it.

Flashlights have light *bulbs* in them.

chemical weathering

[KEM•ih•kuhl WETH•er•ing]

A process by which the chemical makeup of a rock is changed and the rock is broken down.

Water is involved in a lot of *chemical weathering*.

chemical change

[KEM•ih•kuhl CHAYNJ]

A change that results in a new substance; chemical weathering involves chemical changes.

Scientists can produce *chemical changes* in their laboratories.

cleavage

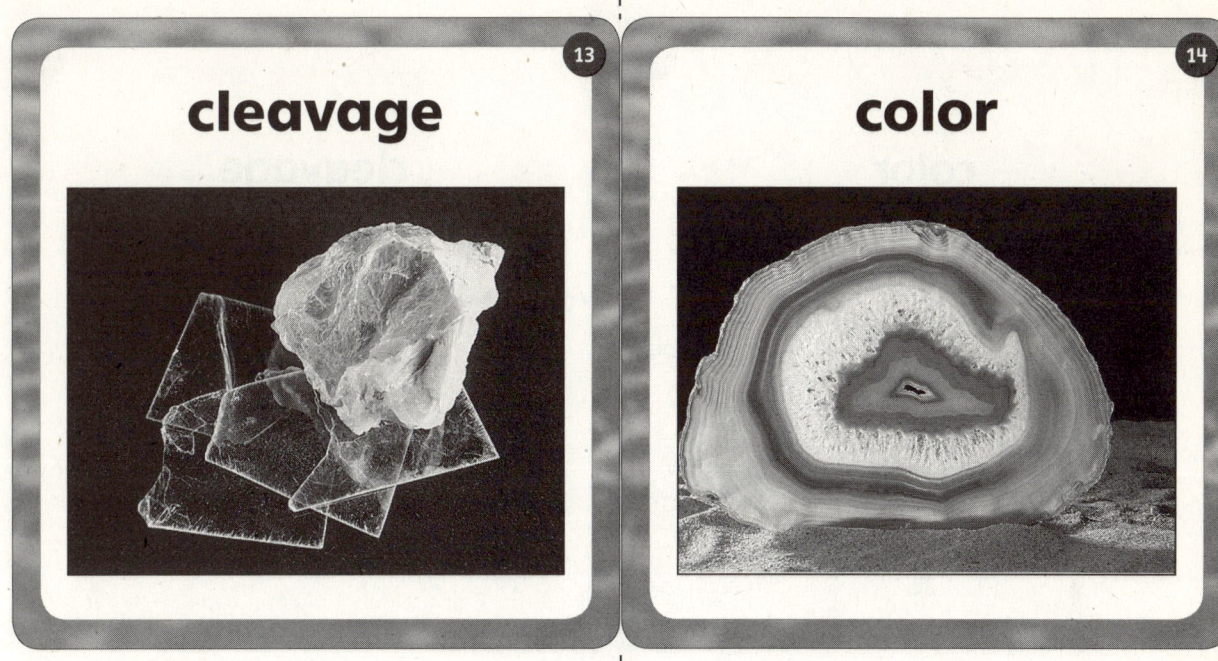

color

community

compass

Vocabulary Cards

Science Content Support CS 119

color
[KUHL•er]

A visual property used to identify minerals.

The *color* of this mineral helps me identify quartz.

cleavage
[KLEEV•ij]

The way some minerals break into pieces with regular shapes.

Mica's *cleavage* is smooth, producing flat sheets.

compass
[KUHM•puhs]

A tool used to determine direction.

This *compass* is pointing north.

community
[kuh•MYOON•uh•tee]

All the populations of organisms living in an environment.

A *community* has many kinds of living things.

competition

consumer

convert

cork

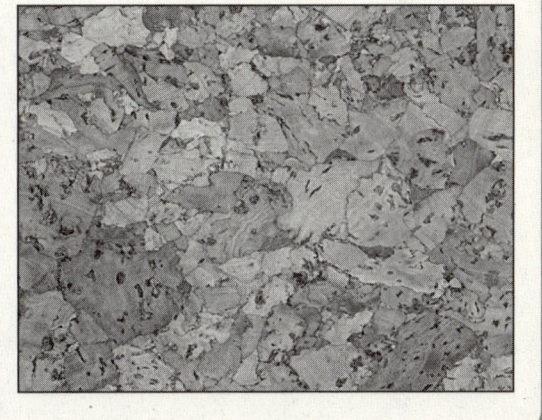

Vocabulary Cards Science Content Support CS 121

consumer
[kuhn•SOOM•er]

A living thing that can't make its own food and must eat other living things.

Animals are *consumers*.

competition
[kahm•puh•TIH•shuhn]

The working of organisms to win limited resources.

A cold winter increases *competition* for food.

cork
[KAWRK]

The lightweight, spongy bark of a kind of oak tree.

You can push pins into the *cork* on a bulletin board.

convert
[kuhn•VERT]

To change.

Pressure *converts* igneous rock to metamorphic rock.

creep

decomposer

deposit

deposition

Vocabulary Cards Science Content Support CS 123

decomposer

[dee•kuhm•POH•zer]

A living thing that feeds on the wastes of plants and animals.

Mushrooms are one kind of *decomposer*.

creep

[KREEP]

The slow movement of soil or rock downhill.

Creep can put soil over mountain roads.

deposition

[deh•puh•ZIH•shuhn]

The dropping of sediment by rivers as they slow down.

Deltas are formed by *deposition*.

deposit

[dee•PAZH•it]

An amount of a mineral in a certain place.

Mines let people get to useful mineral *deposits* in the ground.

dune

earthquake
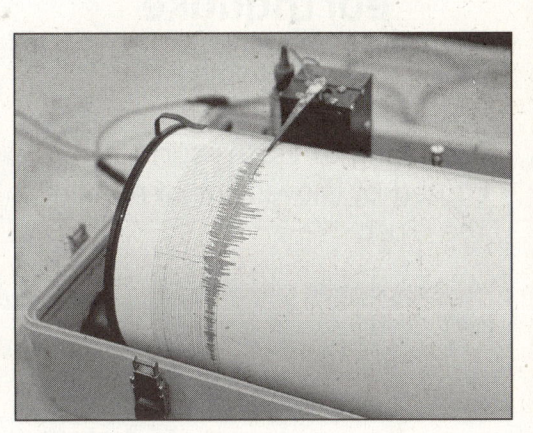

ecology

ecosystem

earthquake

[ERTH•kwayk]

A shaking of Earth's surface, caused by movement of rock in the crust.

This seismograph is recording an *earthquake*.

dune

[DOON]

A large mound of sand piled up by wind.

Dunes can move when the wind blows.

ecosystem

[EE•koh•sis•tuhm]

A community of living things and the community's physical environment.

A pond is one example of an *ecosystem*.

ecology

[ee•KAHL•uh•jee]

The study of ecosystems.

Ecology helps us understand different ecosystems.

electric charge

electric circuit

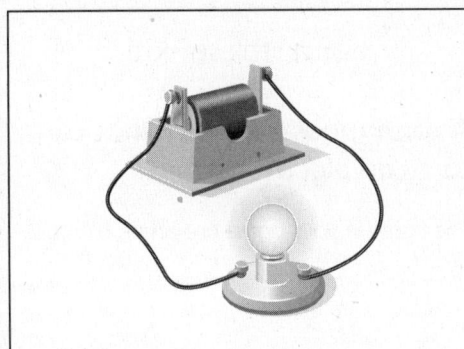

electric current

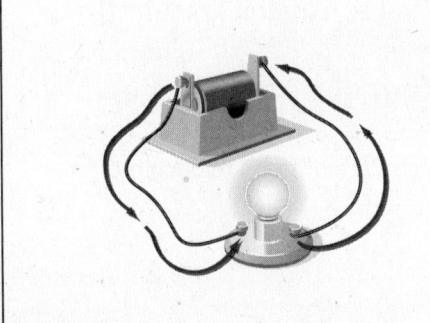

electric field

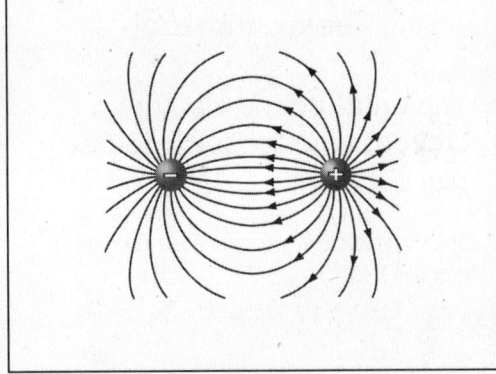

Vocabulary Cards

electric circuit
[ee•LEK•trik SER•kit]

A continuous pathway that can carry an electric current.

The lights in your home are on an *electric circuit*.

electric charge
[ee•LEK•trik CHARJ]

A basic property of the tiny particles that make up matter; it can be positive or negative.

Some particles of matter have *electric charge*.

electric field
[ee•LEK•trik FEELD]

The area around electric charges where electric forces can act.

Strong static electric charges have large *electric fields*.

electric current
[ee•LEK•trik KER•uhnt]

A flow of electric charges.

Electric current flows through a circuit.

electric motor

electricity

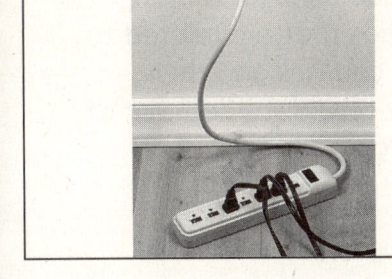

electromagnet

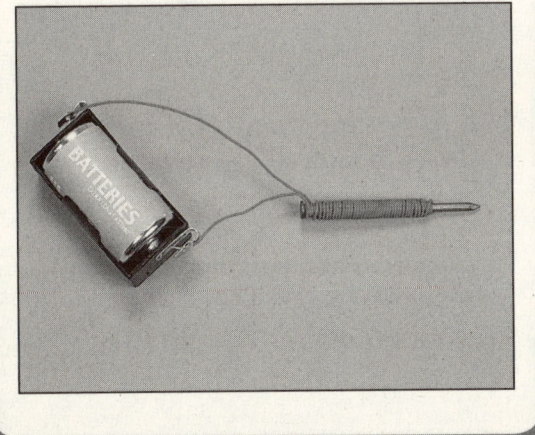

electron

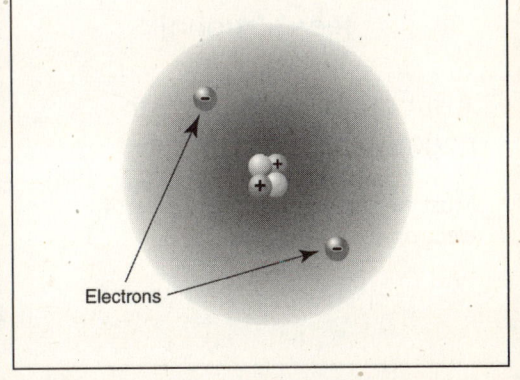

Electrons

Vocabulary Cards

Science Content Support

electricity

[ee•lek•TRIS•ih•tee]

The group of effects and properties that are related to electric charges and their interactions; also the form of energy carried by electric current.

Many things in your home run on electricity.

electric motor

[ee•LEK•trik MOHT•er]

A device that converts electricity to motion.

Some toys have electric motors.

electron

[ee•LEK•trahn]

A subatomic particle with negative electric charge.

Most electric currents are a flow of electrons.

electromagnet

[ee•LEK•troh•mag•nuht]

A device made up of a current-carrying wire coil around an iron core.

You can make an electromagnet with a nail, some wire, and a battery.

energy conversion 37	**erosion** 38
estimate 39	**experiment** 40

Vocabulary Cards

Science Content Support CS 131

erosion
[ee•ROH•zhuhn]

The process of moving sediment from one place to another.

This gully was formed by erosion.

energy conversion
[EN•er•jee kuhn•VER•zhuhn]

A changing of energy from one form to another.

Energy stations perform energy conversions.

experiment
[eks•PAIR•uh•muhnt]

A controlled test of a hypothesis.

An experiment has controlled variables and a dependent variable.

estimate
[ES•tuh•muht]

A careful guess about the amount of something.

When you can't measure something, you might make an estimate.

fault

filament

food chain
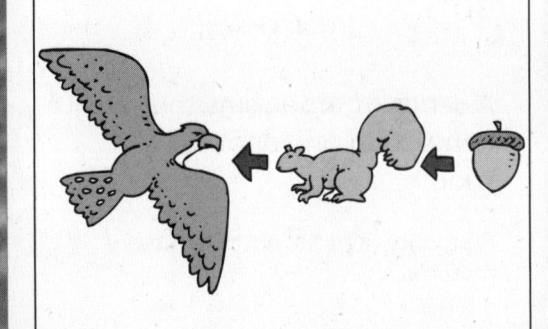

food web

Vocabulary Cards

Science Content Support CS 133

filament
[FIL•uh•muhnt]

The glowing wire coil inside a light bulb.

You can see the glowing *filament* through the clear glass of this light bulb.

fault
[FAWLT]

In Earth's crust, a break along which rocks can move.

The San Andreas *fault* is more than 1300 kilometers (800 mi) long.

food web
[FOOD WEB]

A group of food chains that overlap.

Food webs show interdependence in ecosystems.

food chain
[FOOD CHAYN]

A series of organisms that depend on one another for food.

A producer is at the bottom of every *food chain*.

45 fracture	46 fungus

47 generator	48 habitat

Vocabulary Cards Science Content Support CS 135

fungus
[FUHN•guhs]

An organism that can't make food and can't move about.

Mushrooms are a kind of fungus.

fracture
[FRAK•cher]

A way minerals break if they do not cleave, or break into regular shapes.

Pieces of quartz are often rough because of its curved fracture.

habitat
[HAB•uh•tat]

An environment that meets the needs of an organism.

Different living things need different habitats.

generator
[JEN•uh•ray•ter]

A device that converts other forms of energy into electricity.

People use portable generators when the power is out.

| 49 **hardness** | 50 **heat** |
| 51 **herbivore**  | 52 **hypothesis** 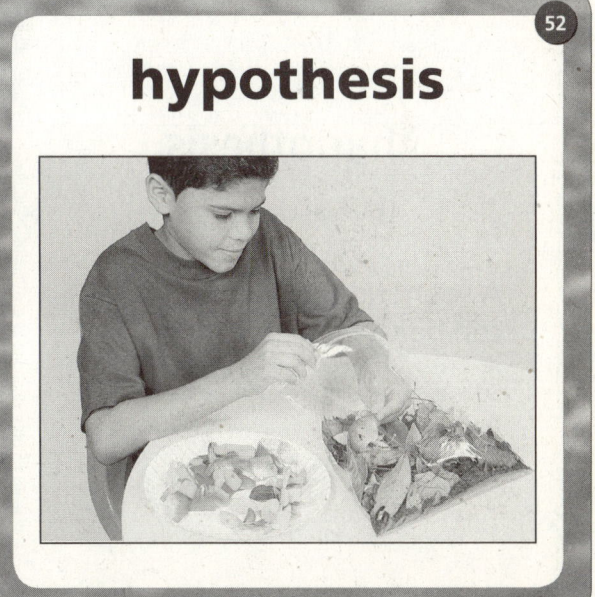 |

Vocabulary Cards **Science Content Support**

heat
[HEET]

The form of energy related to the random motion of the atoms and molecules that make up matter.

These steel bars have been exposed to a lot of *heat*.

hardness
[HARD•nuhs]

The measure of how difficult it is for a mineral to be scratched.

Minerals that can be scratched with a fingernail have a low *hardness*.

hypothesis
[hy•PAHTH•uh•sis]

A scientific explanation that can be tested.

Scientists carry out experiments to test their *hypotheses*.

herbivore
[HER•bih•vawr]

An animal that eats only plants or other producers.

Cows are *herbivores*.

identify	igneous rock 
inference 	insect 

Vocabulary Cards — Science Content Support

igneous rock
[IG•nee•uhs]

Rock that forms when melted rock cools.

Granite is one kind of *igneous* rock.

identify
[eye•DENT•uh•fy]

To recognize and give a scientific name to a sample.

Hardness and color can help you *identify* a mineral.

insect
[IN•sekt]

A type of animal that has three main body parts and six legs.

Ants are one kind of *insect*.

inference
[IN•fer•uhns]

An untested interpretation of observations.

Inferences can help you make hypotheses to test.

insulate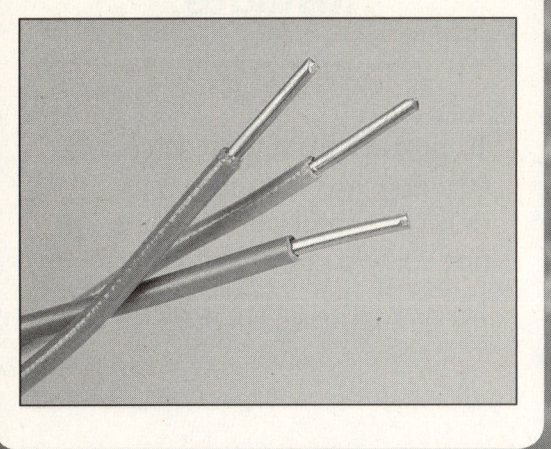

interdependence

interpret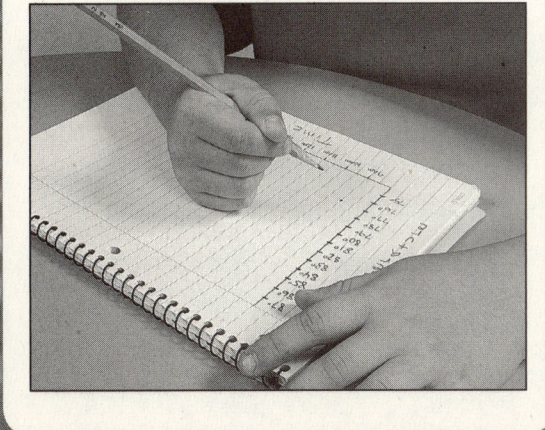

kinetic energy

Vocabulary Cards

interdependence

[in•ter•dee•PEN•duhns]

The dependence of populations of organisms on one another for survival.

Healthy ecosystems show *interdependence* between many kinds of organisms.

insulate

[IN•suh•layt]

To protect from electricity by covering with rubber or another material that does not carry current.

The red, blue, and green plastic *insulates* these wires.

kinetic energy

[kih•NET•ik EN•er•jee]

The energy of motion.

When you move, you have *kinetic energy*.

interpret

[in•TER•pret]

To evaluate evidence or data to draw a conclusion.

Scientists *interpret* their results to explain what happened in their experiments.

landslide	lava
light	lithification

Vocabulary Cards

lava
[LAH•vuh]

Molten rock that flows from a volcano onto Earth's surface.

Lava may flow quickly at first, but it slows down as it cools.

landslide
[LAND•slyd]

The sudden movement of rock and soil downhill.

Landslides can be very dangerous.

lithification
[lith•ih•fuh•KAY•shuhn]

The process of becoming a rock.

Sediment becomes rock through *lithification*.

light
[LYT]

A form of energy that can travel in waves through empty space.

The sun produces *light* energy.

living

luster

magma

magnet

Vocabulary Cards

Science Content Support CS 145

luster

[LUHS•ter]

The brightness and reflecting quality of the surface of a mineral.

Minerals with a metallic *luster* look shiny.

living

[LIV•ing]

Made up of cells and able to react to surroundings and to reproduce.

All animals are *living* things.

magnet

[MAG•nuht]

An object that attracts iron and some other metals.

This *magnet* is strong enough to attract a paper clip even through three sheets of paper.

magma

[MAG•muh]

Melted rock that is beneath Earth's surface.

Magma sometimes hardens into solid rock underground.

magnetic field

magnetic pole

mechanical weathering

metamorphic rock

Vocabulary Cards

Science Content Support CS 147

magnetic pole
[mag•NET•ik POHL]

One of the two areas on a magnet where it exerts the strongest force.

Every magnet has two *magnetic poles*.

magnetic field
[mag•NET•ik FEELD]

The space around a magnet in which magnetic forces can act.

The metal filings show where the *magnetic field* is.

metamorphic rock
[met•uh•MAWR•fik]

Rock that has been changed through high temperature and pressure without being melted.

Gneiss is one kind of *metamorphic* rock.

mechanical weathering
[muh•KAN•ih•kuhl WETH•er•ing]

A process by which rocks are broken down through physical methods.

Wind and water cause a lot of *mechanical weathering*.

microorganism	**microscope** 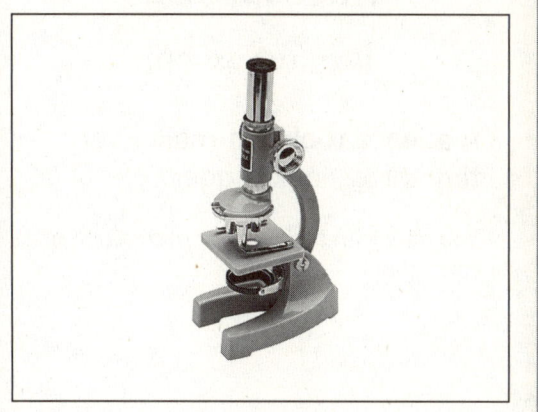
mineral 	**mold** 

Vocabulary Cards — Science Content Support — CS 149

microscope

[MY•kruh•skohp]

A science tool that makes very tiny things look bigger.

You might find this kind of *microscope* at your school.

microorganism

[my•kroh•AWR•guhn•iz•uhm]

An organism that is too small to be seen with the unaided eye.

This picture of a *microorganism* was taken with a microscope.

mold

[MOHLD]

A kind of fungus.

Some kinds of *mold* help people make cheese.

mineral

[MIN•uh•ruhl]

A nonliving solid that occurs naturally and has a repeating structure.

Amethyst is a *mineral*.

molten

motion

mud

mutually dependent

Vocabulary Cards

Science Content Support CS 151

motion

[MO•shuhn]

Any change in position.

This student is in *motion*.

molten

[MOHL•tuhn]

Melted, especially as rock or metal may be.

This *molten* gold is being poured into a mold.

mutually dependent

[MYOO•choo•uhl•ee dee•PEN•duhnt]

Needing each other to survive.

Plants and animals are *mutually dependent*.

mud

[MUHD]

A mix of water and fine sediment.

Some animals and even people use *mud* for building.

needle

niche

nonliving

north-seeking pole

Vocabulary Cards

Science Content Support CS 153

niche

[NICH]

The role a living thing plays in its environment.

Every living thing has a *niche*.

needle

[NEED•uhl]

The small, thin, magnetized pointer in a compass.

The *needle* of a compass points north.

north-seeking pole

[NAWRTH•seek•ing POHL]

The pole of a magnet that moves to point to Earth's north magnetic pole.

The *north-seeking pole* of this magnet points north when the magnet is allowed to turn freely.

nonliving

[nahn•LIV•ing]

Either not made of cells or made of cells but no longer alive.

A bicycle is a *nonliving* thing.

nutrient
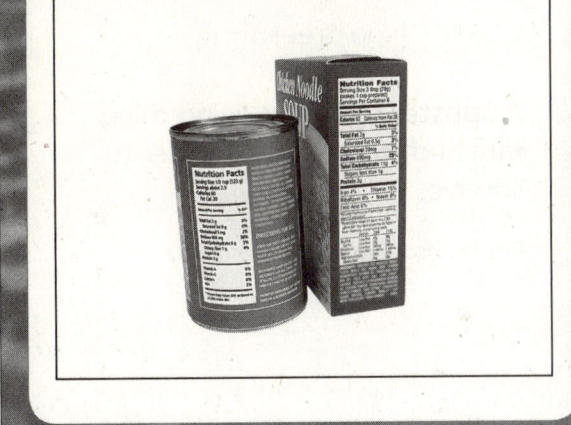

observation

omnivore

parallel circuit
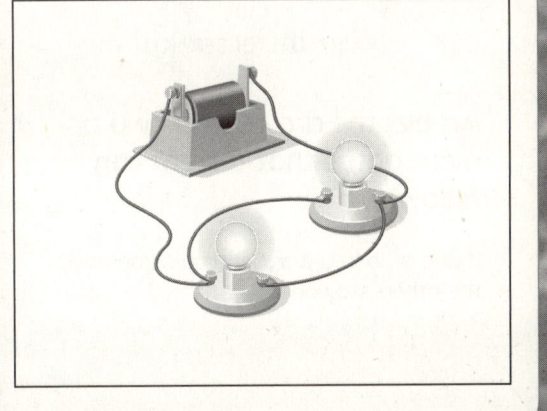

Vocabulary Cards

Science Content Support CS 155

observation
[ahb•zer•VAY•shuhn]

Information that you gather with your senses.

You can make an *observation* with your eyes or ears.

nutrient
[NOO•tree•uhnt]

A substance that a living thing must eat or absorb in order to survive.

Plants absorb *nutrients* from soil.

parallel circuit
[PAIR•uhl•el SER•kit]

An electric circuit with two or more paths that current can follow.

If one bulb in a *parallel circuit* goes out, the others stay on.

omnivore
[AHM•nih•vawr]

An animal that eats both plants and other animals.

A bear is an *omnivore*.

pebble

physical change
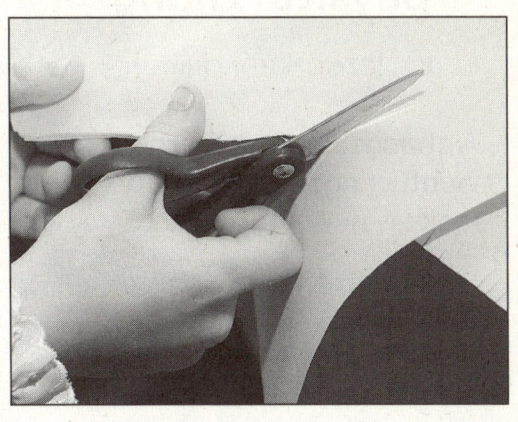

pollinate

population

physical change
[FIZ•ih•kuhl CHAYNJ]

A change, such as mechanical weathering, that does not make a new substance.

The freezing of water is a *physical change*.

pebble
[PEB•uhl]

A smooth, rounded rock that is about 4 to 75 millimeters (0.2 to 3 in.) in diameter.

You can hold several *pebbles* in your hand.

population
[pahp•yoo•LAY•shuhn]

All the individuals of one kind living in the same environment.

Available resources limit animal *populations*.

pollinate
[PAHL•uh•nayt]

To transfer pollen from a male part of a flower to a female part of a flower.

Bees *pollinate* some flowers.

predator

prediction

prey

producer

Vocabulary Cards

Science Content Support CS 159

prediction
[pree·DIK·shuhn]

A statement of what will happen, based on observations and knowledge of cause-and-effect relationships.

Scientists make predictions based on what they already know and careful thought.

predator
[PRED·uh·ter]

A consumer that eats prey.

Cougars are predators.

producer
[proh·DOO·ser]

A living thing, such as a plant, that can make its own food.

Grasses are producers.

prey
[PRAY]

Consumers that are eaten by predators.

Rabbits are one kind of prey.

protist

recycle

relationship

repel

Vocabulary Cards

Science Content Support CS 161

recycle

[ree•sy•kuhl]

To use the material from an object to make new objects.

You can *recycle* your plastic bottles.

protist

[PROH•tist]

A one-celled organism that may share traits with plants or animals.

An amoeba is a *protist*—it has only one cell.

repel

[ree•PEL]

To push away.

The north-seeking poles of two magnets *repel* each other.

relationship

[rih•LAY•shuhn•ship]

A connection between organisms.

Predators have a *relationship* with prey.

resistance

resources

rock

rock cycle

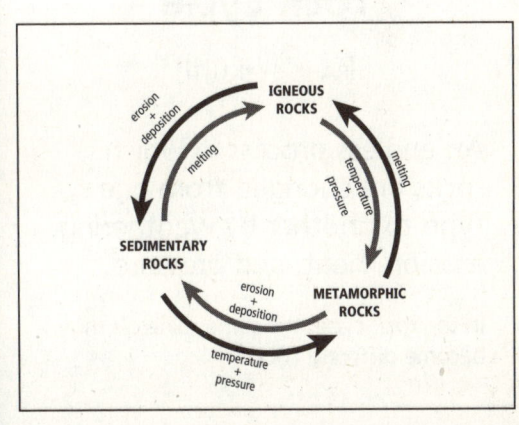

Vocabulary Cards

Science Content Support CS 163

resources

[REEZ•sawr•suhz]

Useful things in the environment, such as air, food, water, and shelter, that animals need to survive.

The amount of available *resources* changes with the seasons.

resistance

[ree•ZIS•tuhns]

The measure of how much a material opposes the flow of electric current.

Most metals have low *resistance*.

rock cycle

[RAHK SY•kuhl]

An endless process in which rocks are changed from one type to another by weathering, erosion, heat, and pressure.

In the *rock cycle*, the same minerals may become different rocks.

rock

[RAHK]

A naturally occurring solid made up of one or more minerals.

Rocks come in all shapes and sizes.

runoff

sand

scale

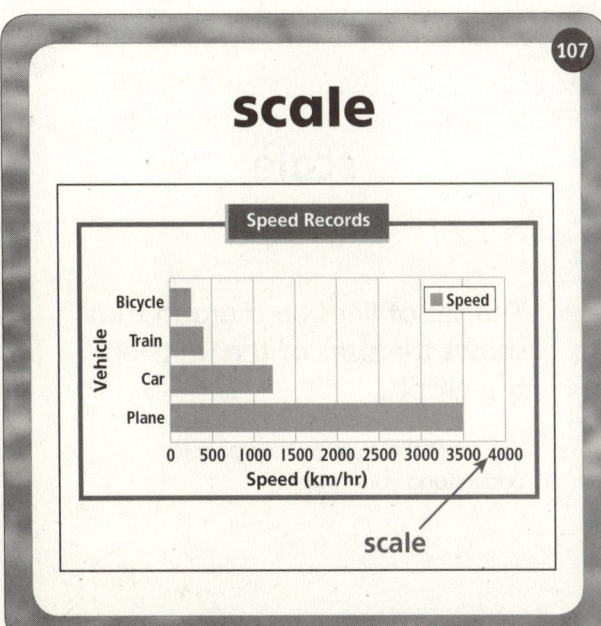

scavenger

sand
[SAND]

Sediment whose pieces are smaller than pebbles and larger than the pieces in silt or clay, about 0.05 to 2 millimeters (0.002 to 0.08 in.) in diameter.

When rocks are broken down into small pieces, they become *sand*.

runoff
[RUHN•awf]

Water that flows over land without sinking in.

Runoff can cause erosion.

scavenger
[SKA•vuhn•jer]

A living thing that feeds on dead organisms.

Scavengers have an important niche in an ecosystem.

scale
[SKAYL]

The set of lines on a graph that shows the sizes of the units on the graph.

You label the *scale* to help people understand your graph.

109 **scientific method** 	110 **sedimentary rock** 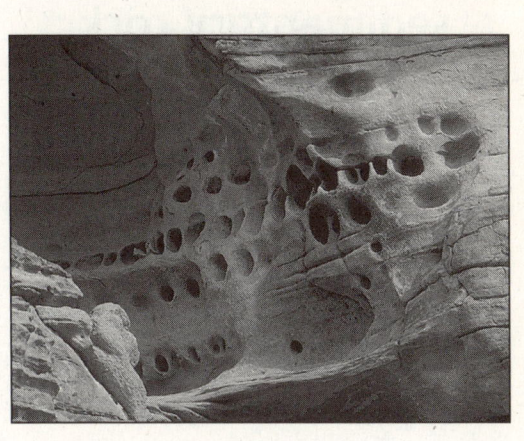
111 **series circuit** 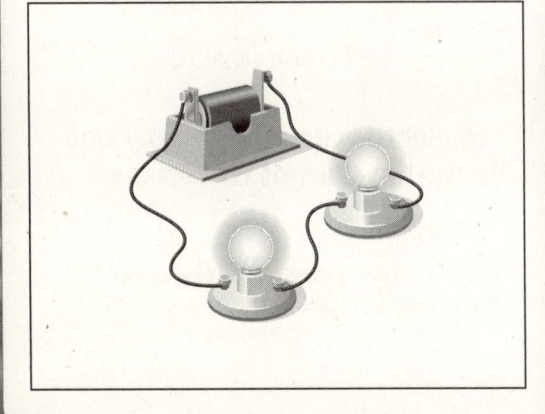	112 **shock** 

Vocabulary Cards Science Content Support CS 167

scientific method

[sy•uhn•TIF•ik METH•uhd]

A series of steps that scientists follow to test hypotheses and find out answers to their science questions.

The students are using the *scientific method* to answer a question about how freezing affects rocks.

sedimentary rock

[sed•uh•MEN•ter•ee]

Rock that forms from sand, mud, and other eroded materials.

Sandstone is one kind of *sedimentary rock*.

series circuit

[SEER•eez SER•kit]

An electric circuit with only one path that current can follow.

If one bulb in a *series circuit* goes out, the others also go out.

shock

[SHAHK]

The painful and possibly dangerous result of electric current flowing through the body.

You can be badly hurt by an *electric shock*.

short circuit 113	**silt** 114 
soil 115 	**solar energy** 116 

Vocabulary Cards

Science Content Support

CS 169

silt

[SILT]

Sediment whose pieces are smaller than sand and larger than the pieces in clay, about 0.002 to 0.05 millimeters (0.00008 to 0.0002 in.) in diameter.

You can find *silt* at the bottom of a river.

short circuit

[SHAWRT SER•kit]

A flaw in a circuit that allows a large current to flow through where it is not wanted.

Short circuits can keep electrical devices from working.

solar energy

[SOHL•er EN•er•jee]

Energy released by the sun.

Solar energy can be changed to electricity.

soil

[SOYL]

A mixture of weathered rock, bits of dead organisms, water, and air.

You can grow plants in *soil*.

south-seeking pole 117	**spring scale** 118 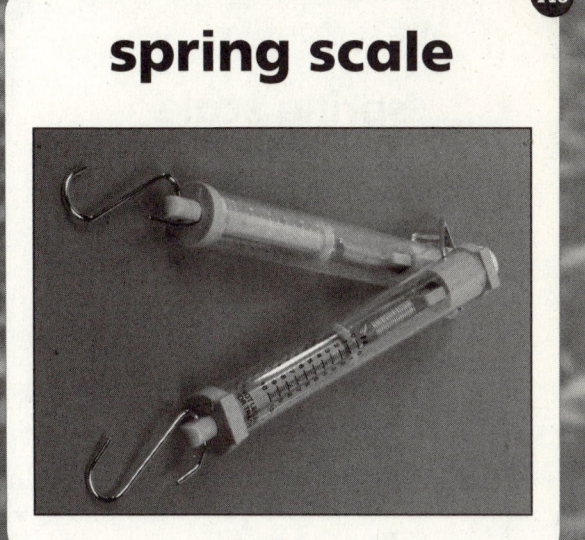
standard measure 119 	**static electricity** 120 

Vocabulary Cards

spring scale

[SPRING SKAYL]

A tool that measures forces, such as weight.

You can find *spring scales* at some grocery stores.

south-seeking pole

[SOWTH•seek•ing POHL]

The pole of a magnet that moves to point to Earth's south magnetic pole.

The *south-seeking pole* of this magnet points south when the magnet is allowed to turn freely.

static electricity

[STAT•ik ee•lek•TRIS•ih•tee]

The buildup of electric charges in one place.

Static electricity might make your hair stand up.

standard measure

[STAND•derd MAZH•er]

An accepted measurement.

A meter is a *standard measure* of length.

Science Content Support — Vocabulary Cards

streak

transport

variable

volcano

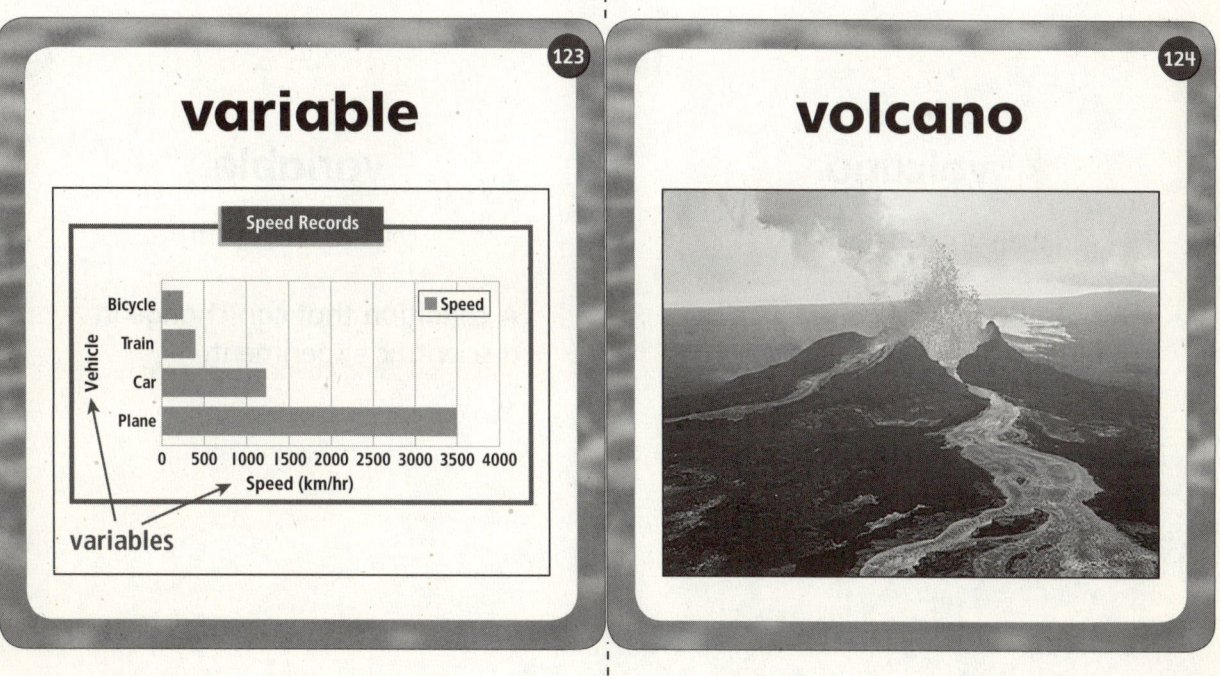

Vocabulary Cards — Science Content Support — CS 173

transport
[TRANS•pawrt]

The movement of sediment from place to place by water.

Transport of sand in the river made those sandbars.

streak
[STREEK]

The mark left by a mineral when it is rubbed across a rough, white tile.

This mineral leaves a brown *streak* on the tile.

volcano
[vahl•KAY•noh]

A mountain that forms as molten rock flows through a crack onto Earth's surface.

The lava from the *volcano* was thick and flowed slowly.

variable
[VAIR•ee•uh•buhl]

A condition that can change in a scientific experiment.

You can control some *variables*.

Science Content Support

Vocabulary Cards

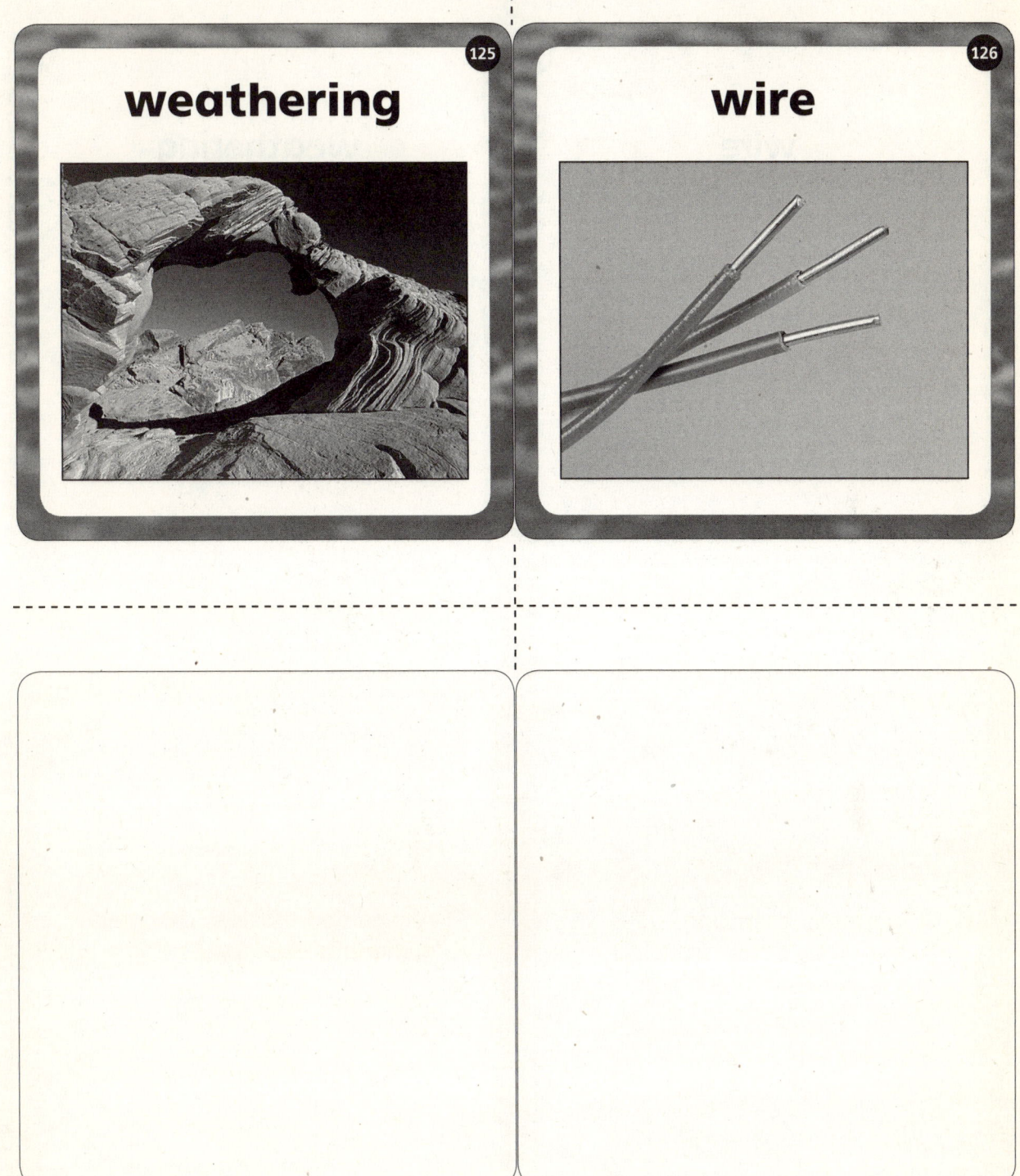

Vocabulary Cards

Science Content Support CS 175

wire

[WYR]

A long, thin piece of metal.

Wires can carry electricity.

weathering

[WETH•er•ing]

A process of breaking down rock.

Weathering has formed a hole in this sandstone, making an arch.